Corrosion In RCC Structures : Repairs & Rehabilitation

OrangeBooks Publication

Smriti Nagar, Bhilai, Chhattisgarh - 490020

Website: **www.orangebooks.in**

First Edition, 2022

ISBN: 978-93-5621-104-9

CORROSION IN RCC STRUCTURES : REPAIRS & REHABILITATION

Dr TARUN GEHLOT, Dr YAGYA SHARMA,
Dr MAYANK DAVE , Dr DEEPANSHU SOLANKI

OrangeBooks Publication

www.orangebooks.in

Preface

The Purpose of this book is to provide a basic course and introduction of corrosion in Reinforced Cement Concrete Structures to students .Various methods of repairs and rehabilitations has been explained with suitable example in this book. One of the challenges today is to repair and strengthen the existing structures subjected to harsh environmental conditions. it is hoped that book will bring about a greater awareness and knowledge to civil engineering students .

This book is divided in to five chapters . chapter 1 consists of general introduction of corrosion and their influence on RCC structures . chapter 2 includes causes of corrosion and details studies and research has done there to cover all possible aspects and mechanisms of corrosions. In chapter 3 repairs and rehabilitations methods and techniques are discussed . all traditional and modern methods are reported there . chapter 4 includes conventional strengthening methods in which materials of repairs and polymer material based methods are discussed. Chapter 5 is structural repair based on extent of damage in which detail studies of cracks in concrete is discussed and repairs method for cracks in RCC structures are discussed .

The authors wishes to express his grateful appreciations to many individuals ,organizations and publications contributing the information.

Dr Tarun Gehlot
Dr Yagya Sharma
Dr Mayank Dave
Dr Deepanshu Solanki

Index

Chapter

1

Introduction

1.1 General

Corrosion is the inevitable process that occurs when refined metals return to their more stable combined forms as oxides, carbonates and sulphides. The corrosion process may be defined as the surface wastage that occurs when metals are exposed to reactive environments. Costs associated with corrosion damage and control can be substantial, being as much as 3.5% of the GDP of some industrial countries.

Reinforced concrete structures have not been immune to the ravages of corrosion despite the protection that concrete provides to embedded steel. reasons for the increasing incidence of corrosion damage to rein- forced concrete structures include the use of deicing salts and calcium chloride set-accelerators, increased construction in aggressive environ- ments, fast-track construction practices, changing cement composition resulting in finer grinding and lower cement contents, lower cover depths and poor construction practice including inadequate supervision.

Reinforcement corrosion is particularly pernicious in that damage may occur rapidly and repairs are invariably expensive. Furthermore by the time visible corrosion damage is noticed, structural integrity may already be compromised. There is currently considerable debate about the merits of the various systems for the repair of reinforcement corrosion.

This book attempts to clarify some of the important issues by drawing on international experience as well as local findings. Ultimately the effectiveness of repair systems should be measured in terms of cost, risk of failure and long-term performance. As such no single system is appropriate for all repairs but will depend on the type of structure, service conditions, level of deterioration and financial constraints of the project .recently many of the Reinforced Concrete structures in contact with water(both seawater and freshwater) have been observed to be suffering from Corrosion Induced damages. The damages observed are concrete cracking, spalling of concrete, efflorescence, water leakages, rust spots, rusted rebars, broken rebars, etc. Some of these defects have been observed within a short span of 3 years from the time of taking these structures in service. In some cases, the damages have been observed after 15-20 years of service. However; the structures are required to be in service for another 20-30 years which means the structures need to be repaired & rehabilitated. Reinforcing concrete structures deteriorate under attack from external elements such as freeze-thaw damage and erosion .Particular concern today is the corrosion of the reinforcing steel, which is affected mainly through carbonation and chloride attack. particular concern today is the alkali silica reaction in the concrete and the corrosion of the reinforcing steel. Bars of which are affected by alkalinity of Port land cement concrete.

Prior to undertaking repair & rehabilitation activities, it is necessary to identify the reasons and extent of damage through detailed condition assessment of the structures. Based on the type and extent of damage, strategies for repair & rehabilitation of the structures need to be developed. The strategies need to include prevention/protection against corrosion for the expected life of the structures. These may include the application of sealers, corrosion inhibitors, surface coatings; polymer modified concrete, cathodic protection/prevention, etc. A life cycle cost analysis is required to be carried out as some of these techniques may appear to be costly but considering the durability & life enhancement of the structures that can be achieved by employing the protection measure, the cost would be more than justified Corrosion of reinforcing steel in concrete is the predominant factor in the premature degradation of reinforced concrete, practical experience and observations suggest that, although many RC structures are seen as badly deteriorated, characterized by mass concrete cracking and spalling, they are still structurally sound. The reason for this is attributed to the nature of the problem; the corrosion product exerts an expansive stress on concrete the tensile strength of which is usually low. It is also partially due to the fact that the safety factors used in structural design for strength are usually larger than those for serviceability since the paramount importance of structural safety. As a result ,corrosion affected RC structures are more prone to cracking , incurring costs of repairs and inconvenience to the public due to interruption. This gives rise to need for proper repair of structures using the cost effective materials available.

Portland cement is made by burning constituents, which include lime in a kiln, and grinding the result to a fine powder. This produces a highly alkaline material, which reacts with water and hardens. When it is added to coarse and fine aggregate and mixed with water the cement combines with the aggregate and hardened to form concrete. The hardening process (hydration reaction) is complex and continues over many months if not years, depending on the amount of water in the mix. Excess calcium hydroxide and other alkaline hydroxides are present in the pores and solution of pH 12.0 to14.0 develops it is this pore network and the solution it contains that are critical to the durability of the concrete. A protective coating of oxides and hydroxides is provided on the surface of the steel reinforcement known as a passive film that protects steel from corrosion when these coatings are destroyed due to severe weathering conditions, leads to very serious problems, which leads to corrosion of reinforced concrete structures.

Usually corrosion is caused due to, the presence of abundant amount of calcium hydroxide and relatively small amount of alkali elements, such as potassium and Sodium in concrete. When concrete structure is often exposed to atmosphere, chloride ions from these will slowly penetrate into the concrete, the chloride ions will eventually reach the steel and then accumulate to beyond a certain concentration level.

Figure 1.1 : RCC structures which are affected by corrosion

We generally think of concrete as a modern building material, yet it is one of the oldest and most durable building materials. Although the Romans experimented with bronze reinforcement, reinforced concrete as we know it today dates from the mid-19th century following the introduction of Portland cement concrete in 1854.

Steel has the advantage of having the tensile strength that concrete lacks, and is highly compatible with its chemical and physical characteristics.

The matching of thermal expansion coefficients is critical to the versatility of reinforced concrete.

Durable concrete is defined as concrete fit for the purpose for which it was intended, under the conditions to which the concrete is expected, and for the expected life during which the concrete is to remain in service.

ACI 201.2R Guide to Durable Concrete – "Durability of hydraulic cement concrete is determined by its ability to resist weathering action, chemical attack, abrasion, or any other process of deterioration"

ACI 201 Deterioration Modes – Freezing & Thawing, Alkali-Aggregate Reaction (AAR), Chemical attack, Corrosion of embedded metal, abrasion

Corrosion is one of the major modes of deterioration of concrete structures and is considered a big threat to the durability of the structures especially for structures in contact with water/seawater.Concrete normally provides reinforcing steel with excellent corrosion protection. The high-alkaline environment in concrete creates a tightly adhering film that passivates the steel and protects it from corrosion. Because of concrete's inherent protective attributes, corrosion of reinforcing steel does not occur in the majority of concrete elements or structures.

Corrosion of steel, however, can occur if the concrete does not resist the ingress of corrosion-causing substances, the structure was not properly designed for the service environment, or the environment is not as anticipated or changes during the service life of the structure Concrete structures located in aggressive environments where high temperatures and high chloride levels exist in both natural soils and waters are subject to premature deterioration from corrosion of the reinforcement.

Traditional methods of corrosion protection, such as concrete admixtures and passive barrier systems, may not be sufficient to provide the level of corrosion control needed for the intended design life. As a solution to this problem, the use of cathodic protection (or cathodic prevention as it is called) at the time of construction is proposed. Although cathodic protection has been used as a rehabilitation method for existing salt-contaminated concrete structures for over 25 years, its application to new reinforced concrete structures is relatively new.

RCC Structures are subject to deterioration through different mechanisms. Although many repair materials have been introduced in the construction industry in the recent past, careful judgment should be exercised by the engineers, while selecting them. The decision should be governed by the results of the in situ testing carried out on the corrosion affected or damaged structure. Periodic maintenance of structures is essential. Each and every problem should be properly analysed and then the appropriate repair methods undertaken. Primary design of the building reflects its performance in long run. Each repair technique is suitable only for the particular application for which it is meant for.

The damaged structures are required to be repaired and Rehabilitated to restore their durability, however; structures affected by corrosion need special treatment to care of corrosion besides restoration of strength. Before undertaking repairs & rehabilitation of damaged structures it is necessary to carryout detailed condition assessment so that suitable remedial measures are taken.

Cracks should not be surface-sealed over corroded reinforcement, without encasing the bars. The methods adopted for repairing spalling and disintegration must be capable of restoring the lost strengthSteel for reinforcement in structures should be provided with proper protection against corrosion. The structures to be built in future should be installed with corrosion resistant reinforcement bars. Various methods for corrosion prevention are there and the right one for the purpose can be chosen.

Cracks should not be surface-sealed over corroded reinforcement, without encasing the bars. The methods adopted for repairing spalling and disintegration must be capable of restoring the lost strength. When complete demolition and total removal and replacement have been ruled out, various strategies are available to the engineer to resolve the integrity of fire damaged structures

The proper method of rehabilitation for different structures should be adopted. The rehabilitation method should be performed without de-stressing the structure. The performance of the repaired or rehabilitated structure is more important such that it should unnecessitate the future rehabilitation for the fore coming years. The need for rehabilitation is due to the improper design of the structures during construction. Just constructing the structures with long-term durable materials can eliminate the rehabilitation processThe materials used for repair and rehabilitation purposes should have high degree of extensibility and should possess long-term durability. As of today the rehabilitation technique by composite materials with fibre-reinforced composites.

The severely deteriorated reinforced concrete frame structures continue to live for years disproving apprehensions of imminent mishap. If carefully nursed through proper rehabilitation techniques based on the fundamental principles elucidated earlier, many a building can be restored to a state of health and vitalityThe modifications in seismic zone map, upgrading in design seismic force and advancement in engineering knowledge has proved that every place in India is predicted to suffer earthquake forces. Hence the existing structures and the structures to be built in future has to seismically designed to resist the lateral forces acting on the structure from the ground motion.

Proper workmanship, low water cement ratio, adequate compaction, proper formwork & strict quality control on the designed ingredients is a must to achieve durable reinforced concrete. In case of development of cracks, treatment with epoxy injection or non shrink cement grout injection should be resorted to at the first sign of distress. If the damage is extensive then treatment with epoxy mortar overlain with cement mortar could be adopted. Cracks are not likely to develop at the junction of old & new concrete if the workmanship &surface preparation is proper .When repairing cracks, do not fill the crack with new concrete or mortar. A brittle overlay should not be used to seal an active crack. The restraints causing the cracks should be relieved, or otherwise the repair must be capable of accommodating future movements.

Reinforced concrete structures require no maintenance or,repair during their service life is gradually being dispelled. It has been said that owners will have to pay for durability at some point in the life of a structure. Inadequate designs with excessive cost-cutting will merely transfer the savings in capital costs to much more expensive repairs at later stage. While accountants may encourage some deferment of capital costs into maintenance, experience suggests that investments in the form of design and construction for durability bring better rewards than allowing for maintenance. Despite this evidence, economic imperatives that attempt to maximise short-term profits, often impact detrimentally on the durability and service life of infrastructural developments.

Repair of reinforced concrete structures needs to be undertaken in a rational manner to guarantee success. An increasing number of repair options are available that must be considered in terms of cost, technical feasibility and reliability. Engineers need to understand all the relevant material, structural and environmental issues associated with concrete repairs in order to make intelligent choices.High quality repairs require a thorough investigation into the causes of deterioration, appropriate repair specifications and competent execution of the repair work. This can only be done when structural investigations are carried out by independent experts, specifications are drawn up by engineers with specialist repair expertise and repairs are undertaken by competent contractors.

Chapter

2

Corrosion In RCC Structures

In order to understand the mechanism behind corrosion of reinforcing steel in concrete, one has to know about the chemical reactions involved in it. In concrete, the presence of abundant amount of calcium hydroxide and relatively small amount of alkali elements, such as sodium and potassium, gives concrete a very high alkalinity with pH of 12 to 13. It is widely accepted that, at the early age of the concert, this high alkalinity results in the transformation of a surface layer the embedded steel to a tightly adhering film that is comprised of an inner dense spinal phase in orientation to the steel substrate and an outer layer of ferric hydroxide. As long as this film is not disturbed, it will keep the steel passive and protected from corrosion.

When a concrete structure is often exposed to atmosphere, chloride ions from these will slowly penetrate into the concrete, mostly through the pores in the hydrated cement paste. The chloride ions will eventually reach the steel and then accumulate to beyond a certain concentration level, at which the protective film is destroyed and the steel begins to corrode, when oxygen and moisture are present in the steel concrete interface.

Once corrosion sets in on the reinforcing steel bars, it proceeds in electrochemical cells formed on the surface of the metal and the electrolyte or solution surrounding the metal, each cell is consists of a pair of electrodes (the anode and its counterpoint, the cathode) on the surface of the metal, a return circuit, and an electrolyte. Basically, on a relatively spot on the metal, the metal undergoes oxidation (ionisation), which is accompanied by production of electrons, and subsequent dissolution. These electrons move through a return circuit, which is a path in the metal itself to reach a relatively catholic spot on the metal, where these electrons are consumed through reactions involving substances found in the electrolyte.

Probably the most frequent cause of damage to reinforced concrete structures is corrosion of reinforcement, and this is usually the result of carbonation of the concrete of the chloride attack. Normally concrete is alkaline (pH12.5 or more) and a passivating layer of oxide quickly forms on the surface of steel embedded in it, if the alkalinity falls below about ph10 the passivating layer is destroyed and, in the presence of oxygen and moisture, the steel will corrode .if chlorides are present in the concrete, passivation is lost at a higher pH value, depending on the chloride ion concentration.

Carbonation occurs as a result of penetration of carbon dioxide from the atmosphere. In the presence of moisture this forms carbonic acid, which neutralizes the alkalinity of the cement matrix. The depth of penetration of carbonation into concrete is proportional to square root of time so that, even if the surface layer of concrete carbonates quickly, the rate of penetration will slow down with increasing depth .the penetration rate depends also on the cement content and permeability of concrete, so that an adequate depth of well compacted cover of good quality concrete will protect the reinforcement for many years. Trouble occurs when the depth of cover is inadequate or its quality is not what it should be. Carbonation penetrates more rapidly into dry than into wet concrete, but both oxygen and moisture, corrosion of steel to occur. Consequently reinforcement corrosion caused by carbonation is found most frequently in concrete exposed to the weather.

With the restriction on chloride contents of materials that are laid down in present day codes of practice, chloride induced corrosion of reinforcement occurs principally in older structures or in those that are exposed to chloride containing materials such as sea water or de-icing salts. It is not possible to specify a limiting chloride content below which corrosion will not occur because a number of factors are involved. A survey by building research establishment has suggested that corrosion is unlikely if the chloride content of concrete is uniformly less than 0.4% by weight of cement and highly probable if it exceeds 1%. The risk of corrosion depends partly on the variability of chloride concentration with in a reinforced concrete element, on the hydroxyl ion concentration and its variability, and the presence of oxygen and moisture without which corrosion will not occur. The chemical composition of cement also has an effect, and chlorides that enters the concrete after it has hardened as, for example, de-icing salts are more harmful than those that are present in the concrete from the start as admixtures or contamination of aggregates. This is so because a proportion of any chloride ions present in freshly mixed concrete will combine with tricalcium aluminates in the cement and will not be available for initiating corrosion.

2.1 Deterioration Of RCC Structures

Unexpected cracking of concrete is a frequent cause of complaints. Cracking can be the result of one or a combination of factors, such as drying shrinkage, thermal contraction, restraint (external or internal) to shortening, sub grade settlement, and applied loads. Cracking can be significantly reduced when the causes are taken into account and preventive steps are utilized. Deterioration of concrete occurs due to one or more of the following mechanisms.

2.2 Corrosion Of Reinforcement Steel

Whatever the source of deterioration and the mechanism of its development, corrosion of embedded reinforcement is recognized as the major problem affecting the durability of concrete structures. Reinforcing steel in good quality concrete does not corrode even if sufficient moisture and oxygen are available. This is due to the spontaneous formation of a thin protective oxide film (passive film) on the steel surface in the highly alkaline pore solution of the concrete. When sufficient chloride ions (from deicing salts or from sea water) have penetrated to the reinforcement or when the pH of the pore solution drops to low values due to carbonation, the protective film is destroyed and the reinforcing steel is depassivated.

When reinforcement corrodes, the corrosion products generally occupy considerably more volume than the steel. The magnitude of this increase in volume varies approximately 7 times the volume of the original material. As a result, the corrosion products produce an internal stress that destroys the neigh bouring concrete under tensile stress. Once corrosion is initiated, it is only a matter of time before the expansive pressures from steel oxidation cause concrete cracking, spalling and de laminations

2.3 Corrosion Process

Steel in concrete is normally protected from corrosion by a passive film of iron oxides on the steel surface resulting from the natural alkaline environment of the concrete. The passive film is chemically stable in the absence of carbonation and chloride ions. The ingress of chloride ions (Cl-) to the level of the steel reinforcing bars destroys the passive film and initiates corrosion. This makes reinforced concrete structures in coastal areas and/or marine environments vulnerable to damage by corrosion of steel reinforcement. Reinforced concrete infrastructures located in cold environments are also susceptible to corrosion damage due to the use of deicing salts.

Carbonation penetrates concrete cover and destroys the passive film by the neutralizing alkalinity of concrete. Once corrosion is initiated, electrochemical reactions occur, leading to the formation of expansive corrosion products that create tensile stresses in the concrete surrounding the corroding steel reinforcing bar. This results in concrete cracking and spalling, which aggravates the progressive damage, thus affecting the durability of the structure.

2.4 Causes of Corrosion

Following are the two most common contributing factors leading to reinforcement corrosion:

i. Localized breakdown of the passive film on the steel by chloride ions called chloride attack.

ii. The general breakdown of passivity by neutralization of the concrete, predominantly by reaction with atmospheric carbon dioxide called carbonation.

2.5 Corrosion: Structural Effects

In the case of concrete structures the first direct effect of the reinforcement corrosion is its section decreases due to the corroding process. Iron oxide (Rust) resulting reinforced concrete structures and their effect induces internal stresses in the concrete, which may lead to cracking or even spalling of concrete.

Accordingly, reduction of structural capacity of reinforced concrete elements affected by rebar corrosion is mainly due to following three main phenomena, which are direct consequence of corrosion:

A. Reduction of section due to corrosion.

B. Reduction of bond strength.

C. Loss of concrete integrity due to cover cracking and spalling.

However corrosion of reinforcing steel can occur by two major situations, they include:

A. Carbonation

B. Chloride contamination.

2.5.1 Carbonation

Carbonation is a process in which carbon dioxide from the atmosphere diffuses through the porous concrete and neutralizes the alkalinity of concrete. The carbonation process will reduce the pH to approximately 8 to 9 in which the oxide film is no longer stable. With adequate supply of oxygen and moisture, corrosion will start. The penetration of concrete structures by carbonation is a slow process, the rate of which is determined by the rate at which carbon dioxide penetrates into the concrete. The rate of penetration primarily depends on the porosity and permeability of the concrete. It is rarely a problem on structures that are built with good quality concrete with adequate depth of cover over the reinforcing steel.Carbon dioxide, which is present in the air at around 0.3 per cent by volume, dissolves in water to form a mildly acidic solution. This forms within the pores of the concrete, here it reacts with the alkaline calcium hydroxide forming insoluble calcium carbonate. The pH value then drops from more than 12 to about 8.5. In the case of carbonation, atmospheric carbon dioxide (CO2) reacts with pore water alkali according to the generalized reaction,

Ca(OH)2 + CO2 converts in to CaCO3 + H2O

It consumes alkalinity and reduces pore water pH to the 8–9 range, where steel is no longer passive. The carbonation process moves as a front through the concrete, on reaching the reinforcing steel, the passive layer decays when the pH value drops below 10.5. If the carbonated front penetrates sufficiently deeply into the concrete to intersect with the concrete reinforcement interface, protection is lost and, since both oxygen and moisture are available, the steel is likely to corrode. The extent of the advance of the carbonation front depends, to a considerable extent, on the porosity and permeability of the concrete and on the conditions of the exposure.

2.5.2 Chloride Contamination

The passivity provided by the alkaline conditions can also be destroyed by the presence of chloride ions, even though a high level of alkalinity remains in the concrete. The chloride ion can locally de-passivate the metal and promote active metal dissolution. Chlorides react with the calcium aluminate and calcium aluminoferrite in the concrete to form insoluble calcium chloro aluminates and calcium chloro-ferrites in which the chloride is bound in non-active form. However, the reaction is never complete and some active soluble chloride always remains in equilibrium in the aqueous phase in the concrete Chloride ions can enter into the concrete from the chloride containing admixtures that are used to accelerate curing or from seawater in marine environment. If the chlorides are present in sufficient quantity, they disrupt the passive film and subject reinforcing steel to corrosion. The levels of chloride required to initiate corrosion are extremely low. Field experience and research have shown that on existing structures subjected to chloride ions, a threshold concentration of about 0.026% (by weight of concrete) is sufficient to break down the passive film and subject the reinforcing steel to corrosion. This equals to 260-ppm chloride. The removal of the passive film from reinforcing steel leads to the corrosion process. Chloride ions within the concrete are usually not distributed uniformly. The steel areas exposed to higher concentrations of chlorides start to corrode, and breakdown of the oxide film eventually occurs. In other areas the steel remains passive. The rate of carbonation in concrete is directly dependent on the water cement ratio of the concrete i.e., higher the ratio the greater is the depth of carbonation in the concrete.

2.6 Mechanism Of Corrosion

The corrosion process that takes place in concrete is electrochemical in nature. Corrosion will result in the flow of electrons between anodic and cathodic sites on the rebar. For corrosion to occur, four basic elements are required:

a. Anode – site where corrosion occurs and current flows from.

b. Cathode – site where no corrosion occurs and current flows to.

c. Electrolyte – a medium capable of conducting electric current by ionic current flow (i.e. soil, water or concrete).

d. Metallic Path – connection between the anode and cathode, which allows the current return and completes the circuit.

The anode is the location on a steel reinforcing bar where corrosion is taking place and metal is being lost. At the anode, iron atoms lose electrons to become iron ions (Fe^{+2}). This oxidation reaction is referred to as the anodic reaction. The cathode is the location on a steel reinforcing bar where metal is not consumed. At the cathode, oxygen in the presence of water, accepts electrons to form hydroxyl ions (OH-). This reduction reaction is referred to as the cathodic reaction. The electrolyte is the medium that facilitates the flow of electrons (electric current) between the anode and the cathode. Concrete, when exposed to wet and dry cycles, has sufficient conductivity to serve as an electrolyte.

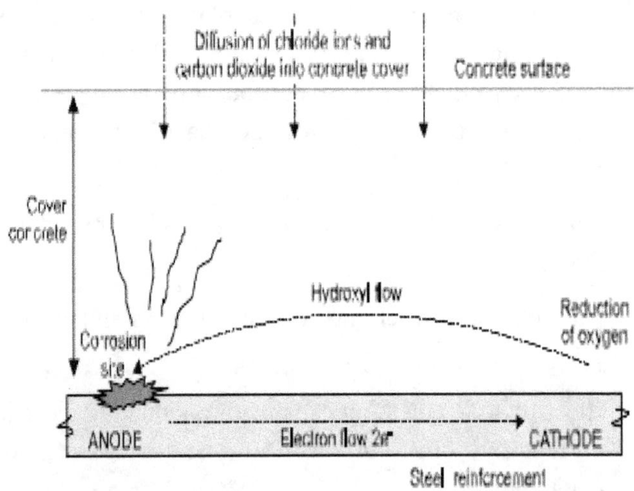

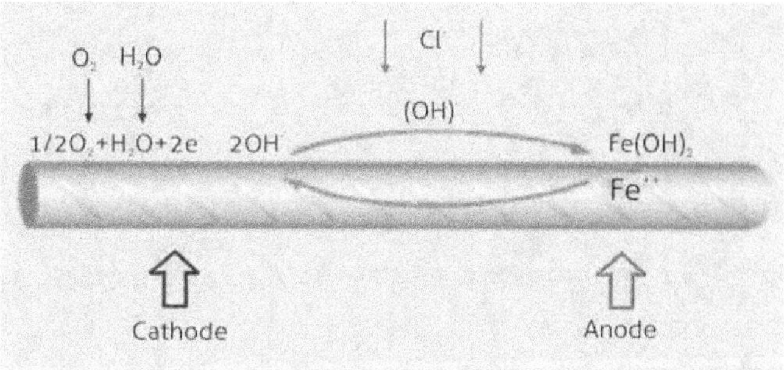

Figure 2.1 : corrosion cell for a steel reinforcing bar embedded in concrete where the anode and the cathode are on the same steel reinforcing bar.

The corrosion of steel in concrete in the presence of oxygen but without chlorides takes place in several steps:

At the anode, iron is oxidized to the ferrous state and releases electrons

Fe oxidized in to $Fe^{2+} + 2e-$

These electrons migrate to the cathode where they combine with water and oxygen to form hydroxyl ions

2e- + H2O + ½ O2 converts in to 2OH- / Fe2+ + 2OH⁻ /Fe(OH)2

In the presence of water and oxygen, the ferrous hydroxide is further oxidized to form Fe2O3

4Fe(OH)2 + O2 + H2O /4Fe(OH)3 oxidized in to /2Fe(OH)3 Fe2O3.2H2O

The anodic and cathodic reactions described above take place at the anode and the cathode, respectively. Currents flow through concrete from the cathode to anode by ion flow and through the reinforcing steel from the anode to the cathode by electron flow.

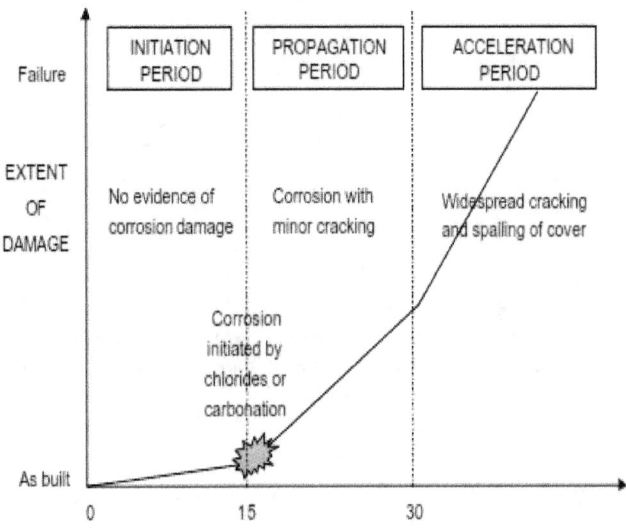

Figure 2.2 : loss of serviceability of corroded reinforced concrete structures via three phase damage model

Two types of corrosion cells develop within concrete: microcells and macrocells. When a microcell develops within the concrete, the anode is very close to the cathode on the same reinforcing bar as shown in the figure. Carbonation-induced corrosion usually occurs on a microcell. When a macrocell develops, the anode and cathode are often well separated as shown in the figure. Chloride-induced corrosion is prone to development of macro cells. The hydroxyl ions combine with the ferrous ions to form ferrous hydroxide.

When carbon dioxide (CO_2) from the atmosphere penetrates concrete and dissolves in the pore solution, carbonic acid is formed. This acid reacts with the alkali in the cement to form carbonates and to lower the pH of the concrete. When the alkalinity reaches a low enough level,

24

the steel reinforcing bar becomes de-passivated and in the presence of sufficient water and oxygen, corrosion is initiated and propagated. The concentration of chlorides in concrete is not uniform due to the heterogeneity of the concrete. The chlorides enter the concrete from the exposed surface. These differences in chloride concentrations establish anodes and cathodes on individual steel bars and result in the formation of micro cells. The corrosion of steel in concrete in the presence of chlorides, but with no oxygen (at the anode), takes place in several steps.

At the anode, iron reacts with chloride ions to form an intermediate soluble iron chloride complex

$Fe + 2Cl^-$ converts in to $(Fe^{2+} + 2Cl^-) + 2e^-$

When the iron–chloride complex diffuses away from the bar to an area with higher pH and concentration of oxygen, it reacts with hydroxyl ions to form $Fe(OH)_2$. This complex reacts with water to form ferrous hydroxide.

$(Fe^{2+} + 2Cl^-) + 2H_2O + 2e^-$ converts in to $Fe(OH)_2 + 2H^+ + 2Cl^-$

The hydrogen ions then combine with electrons to form hydrogen gas

$2H^+ + 2e^-$ converts in to H_2

As in the case of corrosion of steel without chlorides, the ferrous hydroxide, in the presence of water and oxygen, is further oxidized to form Fe_2O_3

$4Fe(OH)_2 + O_2 + H_2O/ 4Fe(OH)_3$ converts in to $/2Fe(OH)_3 Fe_2O_3.2H_2O$

The corrosion products resulting from the corrosion of steel reinforcing bars occupy a volume five to ten times that of the original steel. This increase in volume induces stresses in the concrete that result in cracks, delamination and spalls. If left untreated, the process continues which further accelerates the corrosion process by providing an easy pathway for water and chlorides to reach the steel until the concrete becomes structurally unsound as shown in Fig 2.3

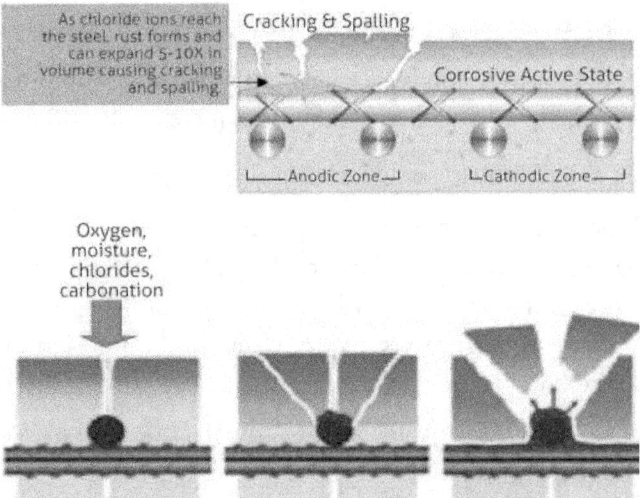

figure 2.3: Corrosion of steel reinforcing bars

The minimum chloride ion concentration needed to initiate corrosion of steel reinforcing bars is also called the chloride threshold. Although the concept of chloride threshold is generally accepted, there is agreement on what the threshold value is. Several factors influence the chloride threshold value (Fig. 2.4)

the composition of the concrete (resistivity), the amount of moisture present, and the atmospheric conditions (temperature and humidity). The threshold concentration depends on the pH level and the concentration of oxygen.

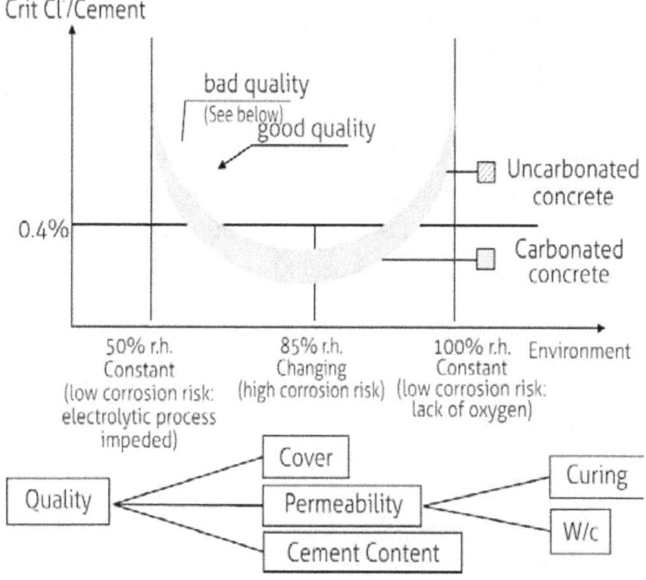

figure 2.4 : factors influence the chloride threshold value

Clearly only pitting and general corrosion represent a threat to the reinforcement and their severity will depend on a number of internal and external factors which need to be assessed when doing a corrosion survey. Internal factors include concrete microstructure, cover depth and moisture condition. External influences such as stray currents and microbial activity may introduce a new dimension into the corrosion system, but are not considered here.

The nature of steel corrosion in concrete depends on local conditions at the surface of the bar. High resistivity concrete with relatively deep covers tends to favour micro-cell corrosion where anode and cathode are close together and cause localized pitting. Conductive concrete contaminated with salt is often able to sustain more widely spaced anode and cathode sites, termed macro-cell corrosion.

Regardless of what concentration of chloride ions is needed to initiate corrosion, an increase in the chloride ion concentration increases the probability that corrosion of steel reinforcing bars will occur.

2.7 Sulphate Attack

Exposure category S – Concrete in contact with soil or water containing deleterious amounts of soluble sulphate

Four Classes

S0 – Very low exposure.

S1 – Structural member in contact with soluble sulphate (moderate) [seawater].

S2 – Structural member in contact with soluble sulphate (severe).

S3 – Structural member in contact with soluble sulphate : high sulphate content (very severe).

2.7.1 ACI Building Code 318: Sulphate Attack on Concrete

a. Negligible attack: When the sulphate content is under 0.1 percent in soil, or under 150 ppm (mg/liter) in water, there shall be no restriction on the cement type and water/cement ratio.

b. Moderate attack: When the sulphate content is 0.1 to 0.2 percent in soil, or 150 to 1500 ppm in water, ASTM Type II portland cement or portland pozzolana or portland slag cement shall be used, with less than an 0.5 water/cement ratio for normal-weight concrete.

c. Severe attack: When the sulphate content is 0.2 to 2.00 percent in soil, or 1500 to 10,000 ppm in water, ASTM Type V portland cement, with less than a 0.45 water/cement ratio, shall be used.

d. Very severe attack: When the sulphate content is over 2 percent in soil, or over 10,000 ppm in water, ASTM Type V cement plus a pozzolanic admixture shall be used, with less than a 0.45 water/cement ratio.

2.8 Effects Of Rebar Corrosion

There are several factors that that affect the corrosion process. The corrosion process leads to several coupled effects such as:

1. Longitudinal cracking of concrete cover due to expansive corrosion products.

2. Steel cross section reduction.

3. The degradation of steel–concrete bond.

As a result of these effects, the service life and the load-bearing capacity of reinforced concrete elements are considerably reduced. The corrosion rate is a key element in determining the time from corrosion initiation to corrosion cracking, which is usually used to predict the functional service life of a corroded RC structure. After corrosion initiation, the corrosion rate depends mainly on the availability of oxygen and moisture at the cathode and on the concrete resistivity, which is mainly affected by the internal moisture content and concrete porosity.

Two major consequences of reinforcement corrosion are commonly observed, cracking and spalling of the cover concrete as a result of expansion of the corrosion product, and a reduction of cross-sectional area of the rebar by pitting (usually only a problem in prestressed con crete structures). Manifestations of corrosion depend on a number of influences   that include:

1. Geometry of the element (large diameter bars at low covers allow easy spalling).

2. Cover depths (deep cover may prevent full oxidation of corrosion product).

3. Moisture   condition   (conductive   electrolytes encourage well-defined macro-cells).

4. Age of structure (rust stains progress to cracking and spalling).

5. Rebar spacing (closely spaced bars in walls and slabs encourage delaminations).

6. Crack distribution (cracks may provide low resistance paths to the reinforcement).

7. Service stresses (corrosion may be accelerated in highly stressed zones).

2.9 Accelerated Corrosion Test

In this test, the specimen consists of a concrete cylinder of 75 mm dia. and 150 mm height, cast with a steel rebar embedded centrally. This test will be a comparative one to assess the relative performance between two specimens. The cured specimens are used for durability study after a period of 28-d after casting.

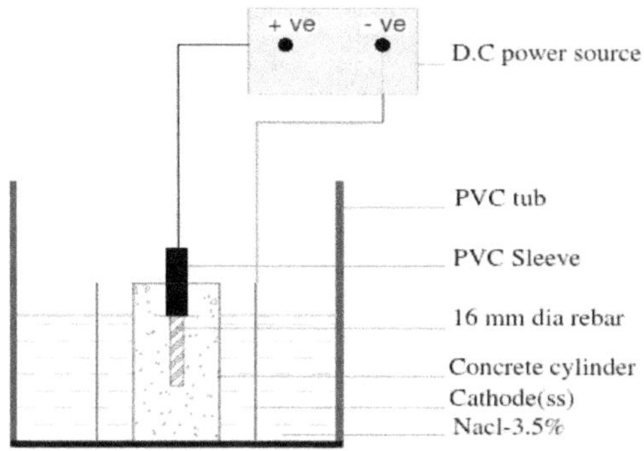

Figure 2.5: Test Set Up for Polarization Experiment

The test set-up consists basically of a non-metallic container in which water combined with 3.5% NaCl is filled to the desired amount. Centrally positioned is the cylindrical concrete sample with bars and a stainless steel plate is held around it. The bar of the concrete cylinder is attached to the anode terminal (+ve) and the stainless steel plate is connected to the cathode terminal (-ve). This arrangement forms an electrochemical cell which acts as anode and as a cathode as stainless steel sheet. A multi-channel machine D.C power pack will make and bind numbers of such cells. The D.C. Power Pack can have a steady voltage of about 5.0 V. Because chloride ions are negative ions, they are drawn to the bar that is the anode of concrete migration. The current response will be measurable through an ammeter for this applied voltage, and this current reply will depend on the overall cell system resistance

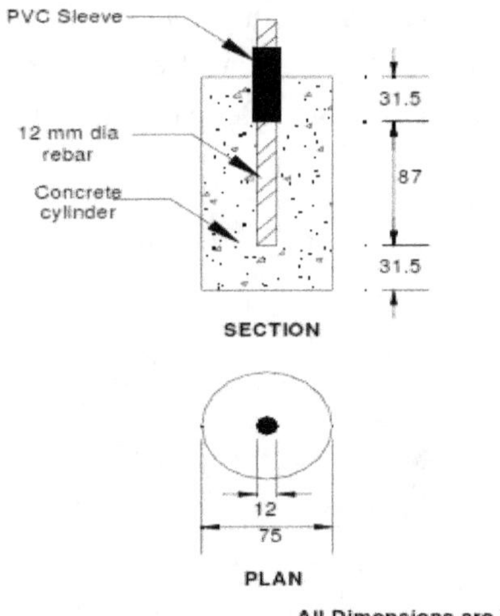

Figure 2.6:Typical Details of Test Specimen

The applied voltage should remain constant and would be time-stamped. Over time, there is an uptick in chloride movement, and as the volume of chloride on the steel reaches its maximum requirement for the type of rebar utilized, de-passivation takes place. If the test advances, the current will increase. Just after chemical assessment, the rebar must be allowed to air-dry before re-shuffling.

Table 2.1 : Parameters Monitored in Polarization Test

Test	Paramete r	Instrument s	Intervals	Duratio n
Polarizatio n Test	Current Reading (mA)	Ammeter	Varies accordin g to concrete type and applied voltage	Varies accordin g to the concrete typeand applied voltage
	Voltage (Volt)	DC Power Supply		

32

2.9 Rapid Chloride Permeability Test (RCPT)

The rapid chloride penetration test (RCPT) apparatus used to test the concrete sample's resistance to the pull of chloride ions is as described in ASTM C1202. The test is carried out by inserting in the sample cells with a diameter of 100 mm and a length of 50 mm and a solution of 3.0 per cent salt and 0.3 N of sodium hydroxide. A 60 V DC voltage is sustained across the test sample, and the charge is reported throughout the sample. A consistency level can be made from the permeability of concrete on the basis of the fee. This is a common fast procedure for assessing the penetrability of chlorine in concrete, introduced by the AASHTO T277 in 1989.

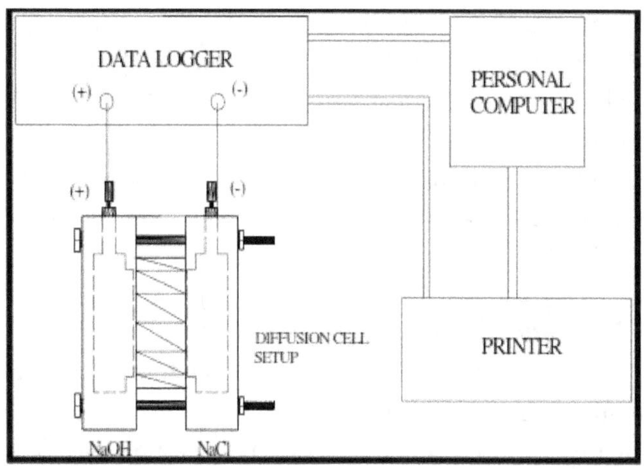

Figure 2.6: Layout of diffusion and RCPT Experiment unit

The concrete specimens with the diameter 100mm and the thickness 50mm have been held in vacuum desiccators to extract all air particles and saturate the concrete specimen completely.

In order to avoid leakage of chemical solutions, the edges were coated with silicon compound. The positive of the terminal was connected with a 0.3M sodium hydroxide solution and the negative terminal with a 2.4M sodium chloride solution and the positive 60-volt supply was connected to the unit. Around 250 ml of solution are stored in each cell.

Two stainless steel electrodes (melts) were added to a specimen with a DC of 60 V and current was measured in coulombs in 30 minutes for a total of 6 hours 30 minutes. As a measure of tolerance of chloride ion concrete to penetration, The total charge passed, in coulombs computed.

As per ASTMC1202 ,If charge passed is greater than 4000 coulombs than chloride ion penetrability remark is high ,similarly for between 2000 to 4000 coulombs remark is moderate ,for 1000 to 2000 coulombs its low remark, for 100 to 1000 coulombs its very low remarks and for less than 100 coulombs penetrability remark is negligible

2.10 Concrete Resistivity Test

Electrical strength of concrete therefore plays an significantfunction in decisiveof corrosion severity Figure 2.7 displays the resistivity measurement. This parameters expressed in terms of "Resistivity" inohm centimeter or in kiloohm cm .

For resistivity greater than 20000 kilo ohm cm than risk of corrosion is negligible whilst for 10000 to 20000 kilo ohm cm corrosion risk is low , for 5000 to 10000 risk is high and for less than 5000 kilo ohm cm corrosion risk is very high

A resistivity review is critical for general monitoring since long-term corrosion can be expected in concrete structures with precisely calculated values below 10,000 kilo ohm cm.

Corrosion ought to be predicted greatly sooner in a structure's life cycle (possibly within 5 years). The theory of concrete resistivity testing is close to that of soil testing. However, when used as concrete, there are a few pitfalls to be aware of.

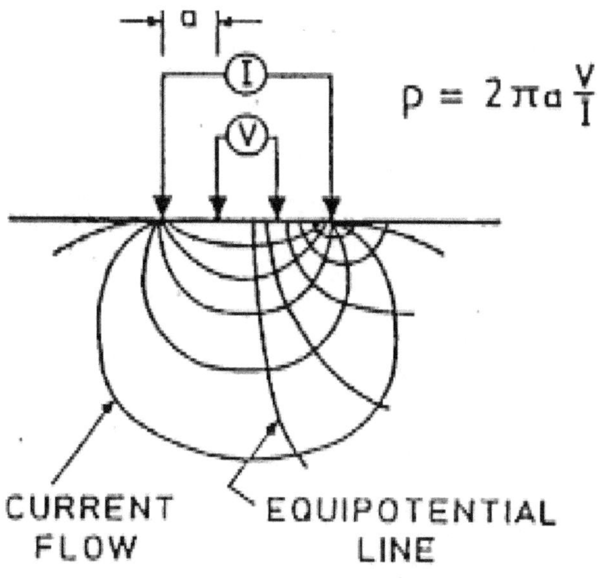

Figure 2.7 : Resistivity Measurements

The method involves using a four-probe methodology where a specific current is applied through two outer probes and the voltage drop between the inner two components is read off, providing for a precise measurement of resistance R using a statistical regression coefficient.

Chapter

3

Repair And Rehabilitation Of R.C.C. Structures

Structure repair and rehabilitating is a process whereby an existing structure is enhanced to increase the probability that the structure will survive for long period of time and also against earthquake forces. This can be accomplished through the addition of new structural elements, the strengthening of existing structural elements, and/or the addition of base isolators. Deterioration of concrete and corrosion of embedded reinforcement structure might make the R.C.C structure structurally deficient. Corrosion can be controlled to some extent by fixing of chloride or protective coating.

Economically, repair and strengthening are often the only viable solution. Differently types of reinforcement require various demolition and surface preparation techniques. Typically, structural deterioration of reinforced concrete members can occur as surface scaling, spalling, cracking, corrosion of reinforcing steel, weathering, post tension losses, deflection beam shortening, volume shrinkage and strength reduction. Moisture, chlorides, carbonation, and chemical attack induce these; freeze thaw disintegration, and sulfate attack, erosion and alkali aggregate reaction. The rehabilitation measures includes epoxy mortar, epoxy bonding coat, epoxy grout, polymer based bonding slurry and mortar, jacketing of columns, shotcreting, epoxy grouting, cement grouting accordingly to the type of distress. The member's load bearing capacity, structural shape and location greatly influence material placement techniques and material selection. The techniques to achieve earthquake resistant design includes; adding base isolators wrapping columns, strengthening footings, adding hinge restrainers, and increasing the width of supports at abutments so that the superstructure will not fall off the support. In repair and rehabilitation process good/sound concrete sharing the load should not be removed for any reason, as is being done today.

3.1 Condition Surveys

A detailed corrosion or condition survey is vital in order to identify the exact caus and extent of deterioration, before repair options are consid- ered. Various diagnostic sheets are given in the Appendix for guidance during condition surveys.

3.1.1 Visual Assessment

Corrosion damage may be identified and defined using a systemat visual survey. Classification of visual evidence of deterioration must be done objectively, following clear guidelines that define damage in terms of appearance, location and cause. Defects may be defined in terms of cracks (caused by corrosion, temperature, shrinkage or fatigue), jointdeficiencies (joint spalls, upward movement, lateral movement, seal damage) surface damage (abrasion, rust stains, delaminations, popouts, spalls), changes in member shape (curling, deflection, settlement, deformation) and textural features (blow holes, honeycombing, sand pockets, segregation). Visual assessment of deterioration can provide useful information when done in a rational, systematic manner but the data may come too late for cost-effective repairs. Rebar corrosion damage is often only fully manifest at the surface after significant deterioration has occurred. Early evidence of distress can sometimes be detected by an experienced engineer before major distress takes place.

3.1.2 De Lamination Survey

A hammer survey or chain drag is a simple method of locating areas of delamination in concrete. Hollow sounding areas can be marked up on the concrete or recorded directly in a survey form. Delamination surveysoften under-estimate the full extent of internal cracking and should not be considered as definitive. Radar and ultrasonic instruments may provide a more sophisticated approach to locating areas of de lamination, particularly at greater depths.

3.1.3 Cover Surveys

Cover surveys are routinely done to locate the position and depth of reinforcement within a concrete structure. Cover meters use an alternating magnetic field to locate steel and any other magnetic material in concrete (note that austenitic stainless steels are non-magnetic). Cover measurements may be unreliable when:

a. Rebar is at deep covers (e.g. covers greater than 80 mm).

b. Measuring regions of closely spaced bars

c. Measuring differing bar types and sizes (unless specifically calibrated).

d. Other magnetic material is nearby (e.g. window frames, wire ties,bolts).

e. To ensure reliable cover depths from a survey, direct measurements of.

f. Rebar depths should be made by exposing a limited number of bars.

g. Calibration can then be made for site specific conditions such as rebar type, concrete and environmental influences.

3.1.4 Chloride Testing

The presence of sufficient chloride at the surface of reinforcement is able to de passivate steel and allow corrosion to occur. Chlorides exist in concrete as both bound and free ions but only free chlorides directly affect corrosion. Measuring free chlorides accurately is extremely difficult and water-soluble chloride tests are unreliable, being strongly affected by the method of sample preparation. Further, bound chlorides may be released into solution under carbonating conditions or by dissolution, making all chlorides in concrete potentially corrosive. Chlorides are therefore most commonly determined as acid soluble or total chlorides.

In accordance with BS 1881 Chloride sampling and determination in concrete is illustrated in Figure 3 and is usually done in the following manner:

1. concrete samples are extracted as either core or drilled powder samplesdepth increments are chosen depending on the cover to steel and thelikely level of chloride contamination (increments are typically between 5 and 25 mm)

2. dry powder samples are digested in concentrated nitric acid to release all chlorides

3. chlorides are analysed using a colorimetric or potentiometric titration

   chloride contents are generally expressed as a percentage by mass of cement

4. chloride profiles may be drawn such that chloride concentrations may be interpolated or extrapolated for any depth (see Figure 3.1) future chloride levels can be estimated from Fick's second law of dif-fusion

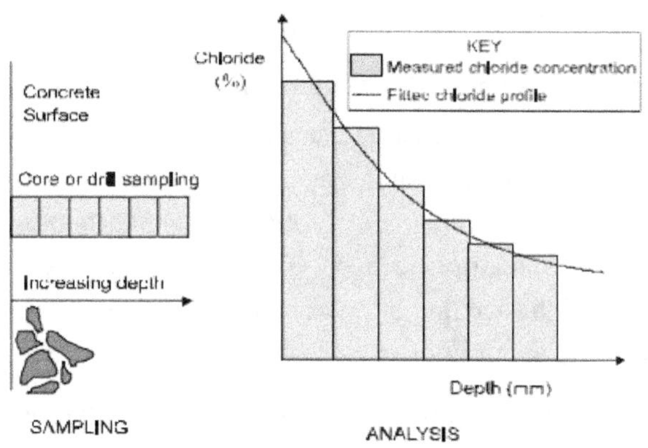

figure 3.1 : chlorides concentration profiles

The corrosion threshold depends on several factors including concrete quality, cover depth, and saturation level of the concrete. The probability of corrosion may be assessed from the following qualitative rating shown in Table 3.1 for acid-soluble chloride contents.

Table 3.1 : Probability of corrosion via Chloride content by mass of cement

Chloride content by mass of cement (%)	Probability of corrosion
< 0.4	Low
0.4 to 1.0	Moderate
> 1.0	High

Limitations of chloride testing of concrete are as follows: presence of chlorides in aggregates may give misleading results chloride contents in cracks and defects cannot be accurately determined slag concretes may be difficult to analyse with colorimetric titration methods relatively large samples are required to allow for the presence of aggregates.

3.1.5 Mechanisms Of Chloride Ion Transportation

Chloride ions can reach the surface of concrete members in different ways, such as by hydrostatic pressure, diffusion and capillary absorption. The diffusion is the most widespread occurrence. The continuing process of liquid and the accumulation of chloride ion are two things that would occur in the concrete member to diffuse.

Permeation is another process whereby chloride ion penetration, induced by pressure gradients can occur.

Take into account a closed volume of concrete with a hydrostatic pressure within the base triggered by a fluid and with a chloride ion, permeations are also possible.

Capillary absorption is another process through which chloride joins. Concrete surfaces also undergo chloride entry through this process when subjected to alternative wetting and drying conditions. In these circumstances, in contact with chloride containing water, the capillary pressure develops due to a moisture gradient and pores absorption of chloride ions. Interestingly, if the concrete content is very weak, chloride ions cannot be captured by this mechanism, since the drying depth is normally very short. It also helps to put the chlorides in concrete and to reduce the distance needed to meet the stainless steel

3.1.6 Chloride Diffusion: A Brief Review

Considering a one-dimensional state for simplicity's sake, Fick's law that governs the diffusion of chloride ions into Chloride states:

$$J = - \text{Deff} \ \frac{dC}{dx}$$

Here, J denotes the flux of chloride ions, Deff denotes the effective diffusion coefficient ,C specifies the concentration of chloride ions and x is a position variable.

Talking in realistic words, the equation written above is useful only after the steady-state setting have been achieved, i.e. the concentration of ions does not change over time. Ficks's second law comes to rescue in unsteady circumstances where focus is shifting over time.

The above equation can be used to arrive at the following equation

$$: \frac{\partial C}{\partial t} = \text{Deff} \quad \frac{\partial^2 y}{\partial x^2}$$

This equation now incorporates the influence that happen due to change in concentration with time (t). On solving the above differential equation and applying the limit condition C(x =0, t >0) = C0 (the surface concentration being constant at C0), the initial condition C(x >0, t=0) = 0 (the initial concentration inside the concrete) and the infinite point situation C(x =¥, t >0) = 0 (distantleft from the surface, the concentration Shall be 0) we get the following solution.:

$$\frac{C(x,t)}{C_0} = 1 - \text{erf}\left[\frac{X}{\sqrt{4D_{eff}*t}}\right]$$

Where erf(y) is the error function, a mathematical construct found in math tables or as a function in common computer spreadsheets

Various considerations that conflict with a clear understanding of diffusion data in functional terms. Firstly, a non-homogenous material (i.e. concrete) with solid and liquid elements, induces the diffusion process. There are obvious differences in the degree and the rate of diffusion in solid portions and in the liquid form. Thus, the diffusion rates are influenced not only by the coefficients of diffusion but also by the physical properties of capillary poles. This influence is usually taken into account in a tacit fashion, but the actual diffusion coefficient in the concrete is here expressly referred to as Deff.

3.1.7 Carbonation Depth

Carbonation depth is measured by spraying fresh concrete with a phenolphthalein indicator solution (1% by mass in ethanol/water solution). Phenolphthalein remains clear where concrete is carbonated but turns pink/purple where concrete is still strongly alkaline (pH > 9.0). Carbonation moves through concrete as a distinct front and reduces the natural alkalinity of concrete from a pH in excess of 12.5 to approxi-

mately 8.3, with a pH level of 10.5 being sufficiently low to de passivate steel. The progress of the carbonation front is shown in Figure 3.2 . Environmental conditions most favourable for carbonation (i.e. 50 – 65 % R.H.) are usually too dry to allow rapid steel corrosion that normally requires humidity levels above 80% R.H. Structures exposed to fluctuations in moisture conditions of the cover concrete, such as may occur during rainy spells, are however vulnerable to carbonation induced corrosion.

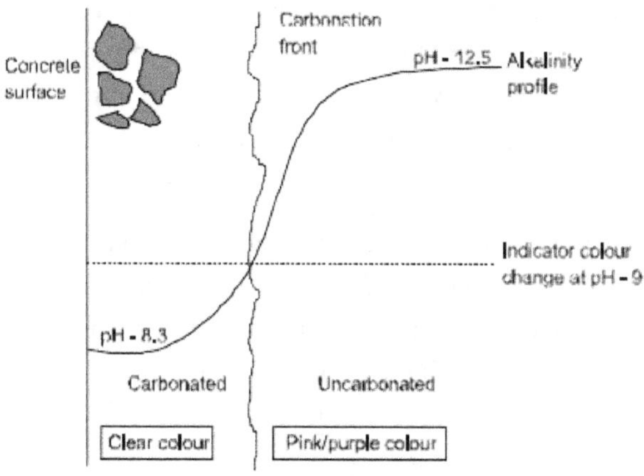

figure 3.2 : Progress of the carbonation front

Limitations to carbonation testing are as follows:

1. Phenolphthalein changes colour at pH 9.0 whereas steel depassivation occurs at pH of approximately 10.5, hence the corrosion risk is slightly under-estimated.

2. Some concretes are dark (e.g. slag concretes) and a distinct colour change is difficult to discern visually phenophthalein may bleach at very high pH levels (e.g. after electrochemical realkalization).

3. Testing must be done on freshly exposed concrete surfaces before atmospheric carbonation occurs.

3.1.8 Rebar Potentials

Chloride-induced corrosion of steel is associated with anodic and cathodic areas along the rebar with consequent changes in electropoten tial of the steel. It is possible to measure these rebar potentials at diffeent points and plot the results in the form of a 'potential map'. Measurement of rebar potentials may determine the thermodynamic risk of corrosion but cannot evaluate the kinetics of the reaction. Rebar potentials are normally determined in accordance with ASTM C876 using a copper/copper sulphate reference electrode connected to a handheld voltmeter.

The qualitative risk of corrosion based on rebar potentials is shown in Table 3.2 Note that the technique is not recommended for carbonation-induced corrosion where clearly defined anodic regions are absent.

Table 3.2 : Qualitative risk of corrosion by Rebar Potential values

Rebar Potential value (mV)	Risk of corrosion
<200	low
200 to 350	Uncertain
>350	High

Rebar potential measurements are relatively quick to perform but have the following limitations:

1. Interpretation of results must be done with caution (preferably by a specialist).

2. Rebar potentials from carbonated concrete are difficult to interpret (the reading is a mixed potential of anodic and cathodic sites).

3. De laminations may disrupt the potential field producing false readings environmental effects will influence potentials (e.g. temperature and humidity).

4. Rebar potentials cannot be directly correlated with corrosion rates stray currents may affect measured potentials.

5. Absolute values are often of lesser importance than differences in rebar potential measured on a structure. A shift of several hundred millivolts over a short distance of 300-500 mm often indicates a high risk of corrosion.

3.1.9 Resistivity

Concrete resistivity controls the rate at which steel corrodes in concrete once favourable conditions for corrosion exist. Resistivity is dependent on the moisture condition of the concrete, on the permeability and interconnectivity of the pore structure, and on the concentration of ionic species in the pore water of concrete such that:

- Poor quality, saturated concrete has low resistivity (e.g. less than 10 kOhm.cm).

- High quality, dry concrete has high resistivity (e.g. greater than 25 kOhm.cm).

3.1.10 Corrosion Rate Measurements

Corrosion rate measurements are the only reliable method of measuring actual corrosion activity in reinforced concrete. A number of sophisticated corrosion monitoring systems are available, based primarily on linear polarization resistance (LPR) principles. These techniques require considerable expertise to operate reliably. Corrosion rate measurements on field structures are most commonly done using galvanostatic LPR techniques with a guard-ring type sensor to confine the area of steel under test. Experience indicates that corrosion rates fluctuate significantly in response to environmental and material influences and single readings are generally unreliable. Table 3.3 shows a qualitative guide for the assessment of corrosion rates of site structures

Table 3.3 : Qualitative assessment of corrosion rate

Corrosion rate (A/cm2)	Qualitative assessment
>10	High
1.0 to 10	Moderate
0.2 to 1.0	Low
< 0.2	Passive

3.2 Repair Strategies

Numerous repair options are available and new technologies continue to make an impact in the field of concrete repairs. The suitability and costeffectiveness o repairs depends on the level of deterioration and specific conditions of the structure

3.2.1 Patch Repairs

Before patch repairs are considered it is important that the distinction between chloride- and carbonation-induced corrosion is appreciated. As a general rule chloride-induced corrosion is far more pernicious and difficult to treat than carbonation-induced corrosion. This often dictates a completely different approach to repairing damage due to the two types of corrosion. Carbonation-induced corrosion causes general corrosion with multiple pitting along the reinforcement. Carbonated concrete tends to have fairly high resistivity that discourages macro-cell formation and allows moderate corrosion rates. Steel exposed to corrosive conditions will therefore show signs of corrosion that can be easily identified (e.g. surface stains, cracking or spalling of concrete). Repairs are generally successful provided all of the corroded reinforcement is treated. Chloride-induced corrosion I characterized by pitting corrosion with distinct anode and cathode sites. The presence of high salt concentrations in the cover concrete means that macro-cell corrosion is possible with relatively large cathodic areas driving localized intense anodes. High corrosion rates can be sustained under such conditions resulting in severe pitting of the reinforcement and damage of the surrounding con crete. Much of the reinforcement may be exposed to corrosive conditions withoutshowing any signs of corrosion, this is particularly noticeable when corroded structures are demolished. Localized patch repairs of areas of corrosion damage are popular dueto their low cost and temporary aesthetic relief. This form of repair has limited success against chloride-induced corrosion as the surrounding concrete may be chloride-contaminated and the reinforcement is therefore still susceptible to corrosion. The patched area of new repair material often causes the formation of incipient anodes adjacent to the repairs as shown in Figure 3.3 These new corrosion sites not only affect the structure but often also undermine the repair leading to accelerated patch fail-ures in as little as two years. Consequently, it is necessary to remove all chloride-contaminated concrete from the vicinity of the reinforcement.

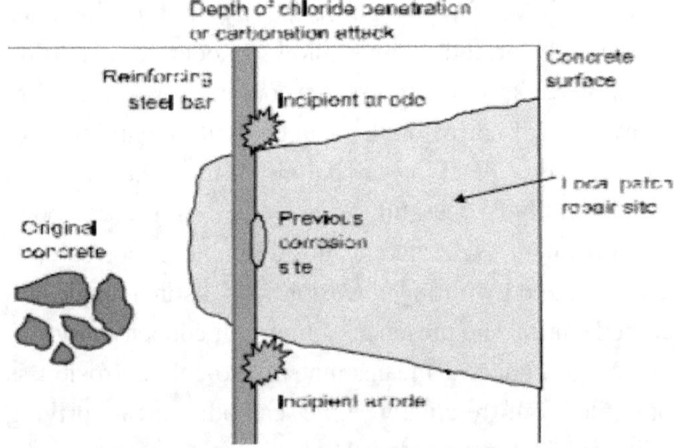

figure 3.3 : formation of incipient anodes after patch repairs

Complete removal of chloride-contaminated concrete, where it is possible should successfully halt corrosion by restoring passivating condi tions to the reinforcement. Mechanical removal of cover concrete is usually done with pneumatic hammer, hydrojetting or milling machines. This form of repair is most successful when treating areas of localized low cover, before significant chloride penetration has occurred. If repair are only considered once corrosion damage is fairly widespread it will be expensive to mechanically remove chloride-contaminated concrete from depths well beyond the reinforcement. Patch repairs consist of the following activities that are briefly described below:-

1. Removal of cracked and delaminated concrete to fully expose the corroded reinforcement.

2. Cleaning of corroded reinforcement and the application of a protective coating to the steel surface (e.g. anti-corrosion epoxy coating or zinc- rich primer coat).

3. Application of repair mortar or micro-concrete to replace the damaged concrete.

4. Possible coating or sealant applied to the entire concrete surface to reduce moisture levels in the concrete.

3.2.2 Coating Systems

A variety of coating and penetrant systems are available that are claimed to be beneficial in repairs of concrete structures. Barrier systems attempt to seal the surface thereby stifling corrosion by restricting oxygen flow to the cathode. In large concrete structures, corrosion control is theoretically unlikely due to the presence of oxygen already in the system. In practice barrier systems are generally ineffective due to the presence of defects in the new coating during application and further damage during service. Such an approach is more likely to promote the formation of differential aeration cells further exacerbating the potential for corrosion. The application of a hydrophobic coating (sometimes referred to as penetrant pore-liners) may be used to reduce the moisture content of concrete and thereby electrolytically stifle the corrosion reaction. The drying action works on the principle that surface capillaries become lined with a hydrophobic coating that repels water molecules during wetting but allows water vapour movement out of the concrete, to facilitate drying. Hydrophobic coatings using silanes and siloxanes are gen erally most effective on uncontaminated concrete, free from cracks and surface defects. The feasibility of such an approach is questionable for marine structures where hi ambient humidity, capillary suction effect and presence of high salt concentrations all interfere with drying.

3.2.3 Migrating Corrosion Inhibitors

A corrosion inhibitor is defined as a chemical substance that reduces the corrosion of metals without a reduction in the concentration of corrosive agents. Corrosion inhibitors work by reducing the rate of the anodic and/or cathodic reactions thereby suppressing the overall corrosion rate. The effectiveness of migrating corrosion inhibitors is generally con- trolled by environmental, material and structural factors, shown in Table 3.4

Table 3.4 : environmental , materials & structural's factors

Likely inhibition	Corrosion condition	Concrete condition	Severity of corrosion
Good	Moderate (carbonation)	Dense concrete with good cover depth	Limited corrosion with minor pitting of steel
moderate	Moderate Chloride level at rebar's	Moderate quality concrete	Moderate corrosion with some pitting
poor	High chloride level at rebar's	Cracked, damaged concrete	Entrenched corrosion with deep pitting

Migrating corrosion inhibitors are generally organic-based materials that move through unsaturated concrete by vapour diffusion. Organic corrosion inhibitors such as amino-alcohols are believed to suppress corrosion by primarily being adsorbed onto the steel surface thereby displacing corrosive ions such as chlorides. The adsorbed organic layer inhibits corrosion by interfering with anodic dissolution of iron while simultaneously disrupting the reduction of oxygen at the cathode.

When assessing the suitability of repairs with migrating corrosion inhibitors, two important issues must first be considered: the likely penetration of the material into the concrete needs to be determined the severity of the corrosive environment at the reinforcement must be quantified

Migrating corrosion inhibitors are designed to move fairly rapidly through partially saturated concretes that allow vapour diffusion. Penetration has however been found to be poor in near-saturated concretes typically found in partially submerged marine structures. This poor penetration performance may be ascribed to high moisture and salt levels that prevent significant vapour diffusion through the concrete. It is critical therefore that satisfactory penetration of corrosion inhibitors is checked before undertaking full-scale repairs.

The performance of migrating corrosion inhibitors in controlling chloride-induced corrosion is largely dependent on chloride levels at the reinforcement. Work done by Rylands indicates that effective inhibition is not possible at chloride levels above 1.0% at the reinforcement can be seen in Figure 3.4 where ribbed steel bars embedded at 25 mm in a grade 40 Portland cement concrete were subjected to wetting and drying cycles with a salt solution for a period of 18 months.

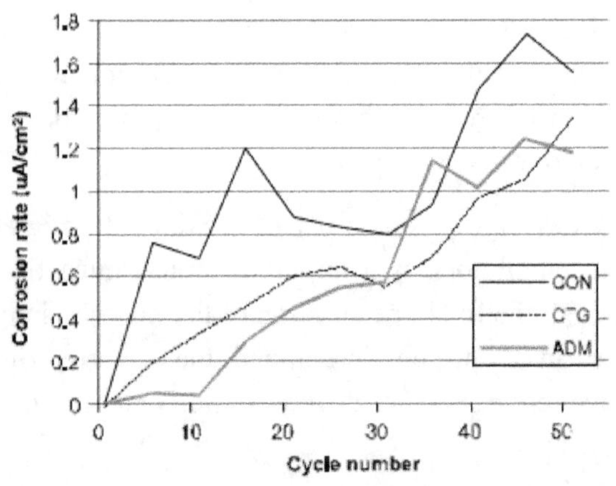

figure 3.4 : corrosion rate verse cycle number

Concrete blocks were either controls (CON) or contained organic corrosion inhibitor, either admixed during casting (ADM) or coated after 30cycles (CTG). The chloride content at the level of the reinforcement was approaching 2% at the time of application of the migrating corrosion inhibitor and resulted in poor inhibition. Better inhibition is possible if treatment is done earlier when chloride contents are lower.   The effectiveness of migrating corrosion inhibitors appears to be enhanced when used in combination with hydrophobic coatings to reduce moisture levels in concrete. This has been noted in both laboratory trials and field monitoring of repairs. Such an approach has also been found to be effective.

3.2.4 Electrochemical Techniques

Corrosion of reinforcement in concrete is an electrochemical process that occurs when embedded steel is de passivated by a reduction in concrete alkalinity or the presence of corrosive ions such as chlorides. Two repair techniques, electrochemical chloride removal and re alkalization, attempt to restore passivating conditions by the temporary application of a strong electric field to the cover concrete region. Re alkalization is the process of restoring the original alkalinity of carbonated

concrete in a non-destructive manner.   The electrochemical treatment consists of placing an anode system and sodium carbonate electrolyte on the concrete surface and applying a high current density (typically 1 A/m 2). The electrical field generates hydroxyl ions at the reinforcement and draws alkalis into the concrete. Alkaline conditions may be restored in the concrete in as little as one to two weeks using the system .Electrochemical chloride removal (ECR) is a more time-consuming and complex technique and its suitability needs to be carefully assessed. Chloride removal is induced by applying a direct current between the reinforcement and an electrode that is placed temporarily onto the out- side of the concrete. The impressed current creates an electric field in the concrete that causes negatively charged ions to migrate from the reinforcement to the external anode. The technique decreases the potential of the reinforcement, increases the hydroxyl ion concentration and decreases the chloride concentration around the steel thereby restoring passivating conditions. Figure 8 shows the basic principles of ECR. The effectiveness of ECR depends on several factors that include the following:

1. Extent of chloride contamination in concrete.

2. Structural configuration including depth and spacing of reinforcement.

3. Applied current density and time of application.

4. Pore solution conductivity and resistance of cover concrete.

5. Presence of cracks, delaminations and defects causing uneven chloride removal.

6. ECR typically takes 4-12 weeks to run at current densities within the normal range of 1-2 A/m2.

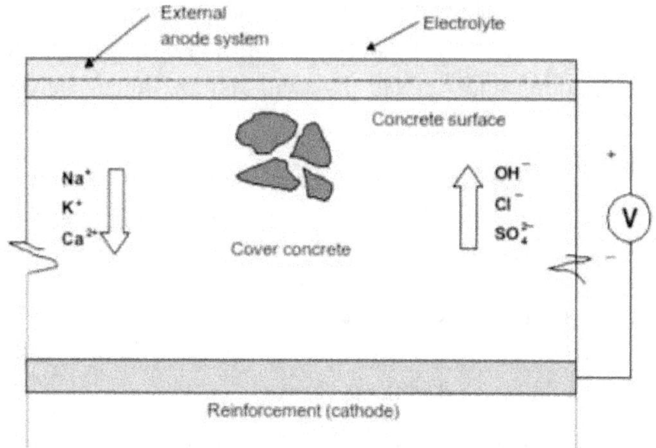

figure 3.5: electro chemical chloride removal method

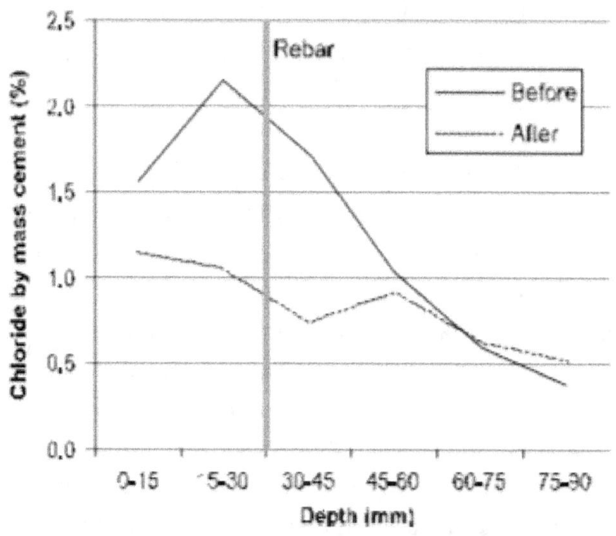

figure 3.6 : chloride by mass cement with depth

Results from ECR trials performed in the laboratory are shown in Figure 3.5 and indicate that complete extraction may take longer than 8 weeks at a current density of 1

A/m2 The feasibility of using ECR depends on several factors such as:• the presence of major cracking, de laminations and defects that will require repair before ECR.

1. Large variations in reinforcement cover that will cause differential chloride extraction and possible short-circuiting.

2. Reactive aggregates requires special precautions to avoid possible alkali silica reaction; lithium salts should be used in these cases.

3. Prestressed concrete structures may be susceptible to hydrogen embrittlemen after ECR; special precautions are needed to eliminate this risk.

4. Temporary power supplies of significant capacity are required during application of ECR.

3.2.5 Cathodic Protection Systems

Cathodic protection systems (CP) have an excellent track record in corrosion control of steel and reinforced concrete structures. The principle of CP is that the electrical potential of the steel reinforcement is artificially decreased by providing an additional anode system at the concrete surface. An external current is required between anode and cathode that diminishes the corrosion rate along embedded reinforcement. The current may be produced either by a sacrificial anode system or using an impressed current from an external power source. Sacrificial anode systems consist of metals higher than steel in the electrochemical series (e.g. zinc). The external anode corrodes preferentially to the steel and supplies electrons to the cathodic steel surface. Sacrificial anode systems are most effective in submerged structures where the concrete is wet and resistivity is low.

Warm temperatures are also generally required for sacrificial CP systems (i.e. above 20 C). CP systems more commonly use an external electrical power source to supply electrons from anode to cathode. The anode is placed near the surface and is connected to the reinforcement through a transformer rectifier that supplies the impressed current (see Figure 3.7). Anodes may be conductive overlays, titanium mesh within a sprayed concrete overlay, discrete anodes or conductive paint systems. Anode systems are usually designed for a minimum service life of 20 years but may last in excess of 50 years.

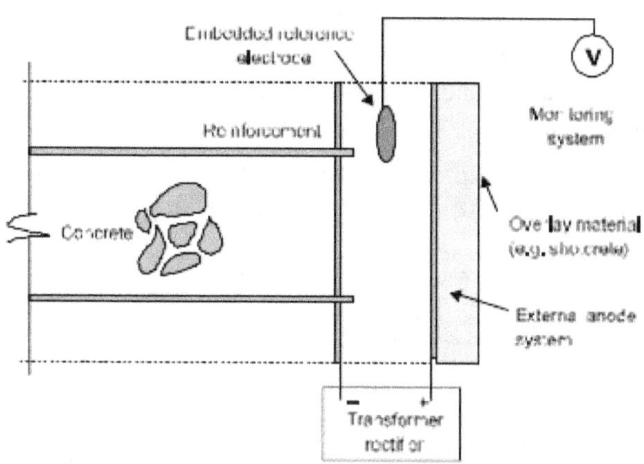

figure 3.7 : transformer rectifier

Before CP repairs are undertaken several factors need to be considered: reinforcement must be electrically continuous concrete cover must be uniformly conductive and free of de laminations alkali reactive aggregates and pre stressing stee need special treatment power must be available to drive the impressed current in the structure CP repair of concrete structures requires a thorough corrosion survey by a specialist and the design needs to be undertaken by a corrosion expert. Reliable CP systems are fully controlled and monitored by a series of embedded sensors in order to ensure optimum performance. This is essential since under or over-protection of the reinforcement may be potentially harmful to the structure or the CP system. Continuous monitoring of CP systems is usually done remotely by modem and the power consumption during operation is extremely small. The first major CP repair of a reinforced concrete structure in South Africa was done at the Simonstown Jetty in 1996 . The structure was almost 80 years old and in an extremely poor condition with widespread chloride-corrosion damage. Several previous patch repairs had failed and the concrete was contaminated with chlorides making conventional repairs unfeasible. An impressed current CP system was installed with metallic ribbon anodes protected within a sprayed concrete overlay.

3.2.6 Demolition/Reconstruction

Deterioration of reinforced concrete structures is often so advanced that demolition and reconstruction becomes viable. This option should only be considered as a last resort since the total cost (capital costs plus loss of service and temporary works) is usually well in excess of repairs costs.Corrosion damage is also generally confined to near-surface regions and engineers often over-estimate the extent of damage to corrosion-damagedb structures. Recent demolition of several bridge-decks along the

Cape coast revealed that actual corrosion damage was less than anticipated. Demolition and reconstruction is often preferred by engineers who have limited repair experience or lack confidence in new repair systems. It is crucial nevertheless that lessons are learnt from the old structure when designing the replacement. Guidance about ensuring durable reinforced concrete structures is given in Monographs 1 and 2.

3.3 Economics Of Repairs

Repairs of reinforced concrete structures damaged by corrosion have often proved to be unsuccessful with further damage occurring after repair. Reasons for the poo performance of repairs include:-

1. Lack of understanding of deterioration processes.

2. Inadequate investigation and testing prior to repairs.

3. Inadequate funds to undertake satisfactory repairs.

4. Ineffective or inappropriate repairs being specified.

5. Poor supervision and implementation of repairs on site.

Repairs are not generally anticipated by owners and funds for repairsare nearly always extremely limited. Economics largely dictate the timingand scale of repairs but unfortunately only short-term costs are oftenther repairs for at least 40-50 years.considered. Whilst corrosion damage is to some degree unique to eachstructure some basic tenets hold for most cases.

1. Performance of the concrete structure prior to treatment often dictates the likely performance after repair. Structures with high levels of damage and rapid rate of deterioration require more substantial repair than those less seriously affected.

2. The timing of treatment is crucial since corrosion rates and damage increase with time. A structure that has been neglected and allowed to reach an advanced level of damage will not respond to 'quick-fix solutions. Conversely a structure that is repaired early enough may be restored to full serviceability relatively cheaply.

3. The effectiveness of treatments in retarding corrosion is not equal and may range from highly effective to detrimental (e.g. cathodic protection versus patch repairs).

Importantly, repairs costs need to be compared in a rational way by comparing cycle costs of the structure. when lifecycle costs are compared, a maintenance-free structural design is cheaper than cutting initial costs and deferring some money for repair and maintenance at a later date (data shown in Table 3.5).

Table 3.5 : Repairs & Maintenance Options

Option	1	2	3	4	5
Original design	60 Mpa 30 % fly ash & 55 mm cover	60 Mpa 30 % fly ash & 30 mm cover	60 Mpa 30 % fly ash & 40 mm cover	60 Mpa 30 % fly ash & 40 mm cover	60 Mpa 30 % fly ash & 75mm cover
Repairs/ Maintenance	none	Surface treatment at 10 year interval	Patch repairs after 20 & 35 years	Cathodic protection after 20 year	Patch repairs after 15,25 and 35 years
Relative cost	1	2	2.3	3	3.5

Notes on repairs options

Option 1: Durability design for maintenance free 40 year service life

Option 2: Based on anticipated life of surface treatment

Option 3 to 4: Based on likely stage at whch spalling damage become excessive

Option 5 : Design reuired by SABS 0100:1992

Repair costs escalate dramatically as deterioration proceed and that repairs should be done as soon as distress is noted This research helped quantify what many engineers had long realized; that durability-based designs are cost-effective in the long-term and that delays in repairs cause an exponential increase in costs. Engineers considering repair of concrete structures do not have thefreedom to change either the original design or the timing of the repairs. Repairs therefore need to be considered on the merits, logistics, costs and risks of the many options that are available to rehabilitate the structure. To illustrate some of the issues that need to be considered, a practical example is given below.

Repair example

A 60-year old bridge structure is in need of major repairs arising from widespread corrosion damage. The bridge spans a tidal estuary with direct exposure to seawater splash and spray action. Concrete is heavily contaminated with salt and chloride levels at the reinforcement are around 1.0% by mass of cement. Damage in the form of cracking, spalling and de laminations are widespread over much of the structure and are the result of chloride-induced corrosion. Urgent repairs are essential to restore full serviceability to the bridge. Rough estimates of service life of the various options are based on recent trends.

Whilst the projected performance of the various repairs is a subjective assessment, the figures serve to illustrate the many issues that need to be considered when costing repairs. For the purposes of costing the repair options, the following assumptions are made:-

1. Un escalated 2001 costs are used due to uncertainties about future discount, inflation and tax rates.

2. Site establishment costs are fixed at R 250 000 for each repair option.

3. Total area of concrete under repair is 2000 m2.

4. Unit rates for repair include allowance for labour, materials, access and supervision.

5. Repairs are focused on chloride-induced corrosion damage only.

The following repair options are considered for the bridge.

A. Localized repairs of corrosion-damaged areas with only cosmetic consequences. Assuming 15% of the structure requires patching and that concrete is only broken back to the reinforcement, a unit rate of R 250/m2 is used. Given the limited nature of the repairs and the likelihood of incipient anode formation an effective life of 8 years is considered possible.

B. More extensive mechanical break-outs and patching are done with all corroded reinforcement being exposed, cleaned and a good quality repair material used for patching. Approximately 30% of the structure is treated at a unit rate of R 280/m2 Despite the effort made to repair the structure, corrosive conditions still exist at the reinforcement and further corrosion damage limits the effective life to 12 years before more repairs must be considered.

C. Conventional corrosion repairs are done but a migrating corrosion inhibitor is applied to the repaired concrete surface together with a hydrophobic coating (silane/siloxane). Mechanical breakout is limited to damaged areas of concrete and not all corrosion on reinforcement is removed resulting in a unit rate of R300/m2 This includes the cost of the migrating corrosion inhibitor and coating at R40/m2. The chloride level at the reinforcement (1.0%) is at the upper level for corrosion inhibitor performance resulting in an effective service life of only 15 years.

D. Electrochemical chloride extraction is applied to the concrete to remove chloride from around the steel. The cost of the system is approximately R750/m for a six week application and includes repair to damaged concrete. Unfortunately not all the chloride is removed from the concrete resulting in an effective service life of 25 years.

E. Cathodic protection is applied to the structure to protect the embedded reinforcement. The cost of the system is R900/m2 at installation and a nominal maintenance and monitoring fee of R5000 per year The anode system is designed to last 50 years thereby dictating the effective life of the system.

Present value costs for the various options are shown in Table 3.6. From these findings it is clear that initial repair costs and total repair costs over 40 years vary significantly. Option A is most cost-effective when only short-term costs are considered but most expensive in the longer-term. For a structure that only has to last another 20 years, option C may be preferable whereas for 40 years further service, option E is most economical for the hypothetical example.

Table 3.6 : Options for repairs costs

Timing	Option A	Option B	Option C	Option D	Option E
initial	0.75	0.81	0.85	1.75	2.05
20 years	2.25	2.43	1.70	1.75	2.15
40 years	3.75	3.24	2.49	3.50	2.25

3.4 Surface Preparation For Application Of Patch Repairs, Sealers And Coatings In Concrete Repair

The main purpose of surface preparation is to provide maximum coating adhesion and increase the surface area by increasing the roughness of the surface. Achieving an adequate lasting bond between repair materials and existing concrete is a critical requirement for durable concrete repair. Good surface preparation using proper concrete removal methods and workmanship is the key element in a long-lasting concrete repair technique.

3.4.1 Grouting Process

Grouting is the process of placing a material into cavities in concrete or masonry structures for the purpose of increasing the load bearing capacity of a structure, restoring the monolithic nature of structural member, filing voids around pre cast connections and steel base plates, providing fire stops, stopping leakages, placing adhesives and soil stabilization. Methods of application normally used include: hand pumps, piston pumps, single and plural component pumps, gravity and dry packing placement, micro capsules and single co Grouting of cracks can be performed in the same manner as the injection of an epoxy, and this technique has the same areas of applications and limitations. However, the use of an epoxy is the better solution except where the considerations of fire resistance or cold weather prevent such use, in which case grouting is the comparable alternative. The procedure is similar to other grouting methods and consists of cleaning the concrete along the crack; installing built-up seats at intervals along the crack, sealing the crack between the seats with a cement paint or grout, flushing the crack to clean it and test the seal; and then grouting the whole. The grout itself is high early strength Portland cement. An alternative and better method, where it can be performed, is to drill down the length of the crack and grout it so as to form a key. The grout key functions to prevent relative transverse movements of the sections of concrete adjacent to the crack. However, this technique is applicable only where the cracks run in a reasonably straight line and are accessible at one end. The drilled hole should preferably be 2 or 3 inches in diameter and flushed to clean out the crack and permit better penetration of the grout component pressurized cartons.

3.4.2 Guniting Process

Guniting is an effective technique, which has been extensively used in the rehabilitation of structurally distressed RC members. There have been cases of heavy rusting of the mesh in the form of powder or in the form of a sheet coming out. De-stressing before restoration is possible only in the case of overhead tanks which can be restored when the tanks are empty. The guniting technique suffers from other drawbacks like dust and noise nuisance.Gunite is also known as shotcrete or pneumatically applied mortar. Gunite is used for the restoration of concrete surfaces where the deterioration is relatively shallow. It can be used on vertical and overhead, as well as on horizontal surfaces and is particularly useful for restoring surfaces spa lied due to corrosion of reinforcement. Gunite is a mixture of Portland cement, sand and water, shot into the place by compressed air. In structural applications, the sand and cement are mixed dry in a mixing chamber, and the dry mixture is then transferred by air pressure along a pipe or hose to a nozzle, where it is forcibly projected on to the surface to be coated. Water is added to the mixture by passing it through a spray injected at the nozzle. The flow of water at the nozzle can be controlled to give a mix of desired stiffness, which will adhere to the surface against which it is projected. The existing surface must be made rough to afford a good keying effect. Anchor bolts tying the new work to the old concrete are essential. A layer of galvanized welded wire mesh can also be provided. The mesh is connected to the hook bolts and the existing reinforcing bars, with galvanized tie wire. Incompatibility of gunite with the old concrete is a major problem, and the keying effects of the rough surface and the doweling effects of the anchor bolts are necessary to assure interaction between the two materials. Sand for gunite should be uniformly graded, as for conventional concrete. Hard particles are desirable, since there is a tendency to grind and crumble the grains as they pass through the discharge hose. The sand should contain 3 to 5 percent moisture for efficient operation of the equipment.

Because of impact, a certain amount of the material being projected against the surface to be coated will bounce off. This material is known as rebound. It consists primarily of the coarse sand particles and has a much smaller cement content than the mix, as projected from the nozzle. The occurrence of some rebound is unavoidable, and it amounts to 20 to 30 percent. Because of the rebound, the cement content of the mortar, in place, will be substantially greater than that of the materials as fed to the mixer. Thus, if a 1:3 mix is desired in place, a 1:4 mix fed into the mixer might be adequately rich. It is difficult to mention mix proportions for gunite because of uncertainties and variations in the amount of rebound, because of difficulties in making test cylinders, and because there is little control over the water-cement ratio of the material in place. In practice, it is usual to use as much water as possible without causing fallouts, as this minimizes the rebound. Roughly, this might mean a water cement ratio of 0.5 to 0.6. Starting with this assumption and assuming a 20 to 30 percent rebound, the proportions required for the mix, as fed to the mixer, can be approximated for any desired strength. In practice, a 1:3 or 1:3.5 mix, by volume, is fed into the mixer.

The sand and cement shall be thoroughly mixed in dry state. The time of mixing shall not be less than 1.5 minutes. The nozzle shall be held between 2ft and 4ft from the surface to be coated, and held in such a position that the flow of material will strike it as near to a right angle as possible. In shooting vertical or sloped surfaces, the placing shall be started at the bottom and carried up. In such cases, the mortar shall be placed in layers of such thickness that the weight of the plastic mass does not cause it to sag. When more than one layer is to be used to complete the final thickness of the work, the delay between applications of successive layers shall be ample to prevent sagging or fallout of the mass, but not so long that the underlying layer has completely set and developed a glaze coating. It is observed that 30 minutes to 1 hour is usually a proper interval. Before any gunite is applied, care shall be taken to remove any sand or rebound clinging to the surfaces. No gunite shall be applied to a surface on which there is running or free water. When shooting around reinforcing bars or anchor bolts, the nozzle shall be moved from side to side and angled to place the gunite back of the rod. At the end of the day's work, or at similar stopping periods, the gunite shall be tapered to a thin edge. Before shooting the adjacent section, this tapered portion shall be thoroughly cleaned and wetted. Operations shall be suspended when wind velocity is such that it blows away the spray from the nozzle and prevents proper control of the consistency. As soon as dry patches begin to appear on the surface of the newly placed mortar, curing by use of a water spray or application of two coats of an approved sealing compound shall be commenced. Minimum curing period shall be 7 days.

The following points need to be kept in mind for better results of guniting:

a. Coating of existing as well as new bars by zinc rich epoxy primer to guard against corrosion.

b. Mesh reinforcement is not advised.

3.4.3 Application Of Epoxy Resins To Strengthen The Structural Member With External Reinforcement

The prolonging service life and enhancing the load-carrying capacity of reinforced concrete (RC) structures calls for new approaches in addition to the well-established techniques. The recent development in the field of structural adhesives, particularly the ones that are based on epoxy resins, has led to the development of new technique of strengthening RC structural by adhesively-bonded external plates. In these methods of strengthening, an epoxy adhesive (normally consisting of two components - a resin and a hardener) is used to bond steel plates to overstressed regions of RC members. Normally, the steel plates are located in the tension zone of concrete to enhance the flexural capacity. The plates can also be placed in the compression and shear regions for enhancing the axial and shear-capacities of the RC structural elements. As the adhesive provides a continuous shear connection between the RC member and the external plates, a concrete-adhesive-steel composite structural member is developed to cater for the additional live load effects on the structures.In these methods of strengthening, an epoxy adhesive normally consisting of two components- a resin and a hardener is used to bond steel plates to overstressed regions of RC members. Normally, the steel plates are located in the tension zone of concrete to enhance the flexural capacity. The plates can also be placed in the compression and shear regions for enhancing the axial and shear- capacities of the RC structural elements. As adhesive provides a continuous shear connection between the RC member and the external plates, a concrete-adhesive-steel composite structural member is developed to cater for the additional live load effects on the structures.

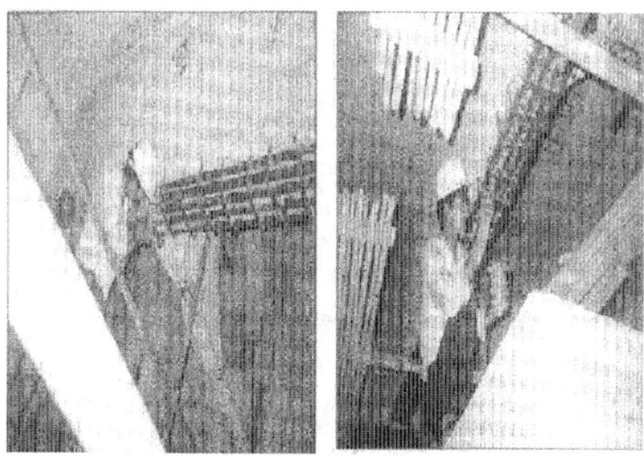

Fig 3.7: Application of protective coating   on steel

Chapter

4

Conventional Strengthening Methods

4.1 Section Enlargement/Jacketing

In this method the entire height of the column section is increased and a cage of additional main reinforcement bars with shear stirrups is provided right from the foundation as per requirement of additional load, etc. However there are many instances where the column section is increased with additional reinforcement bars only on one face, and that too starting from the floor slab level of particular floor and only up to height of deterioration of the column. The enlargement should be bonded to existing concrete to produce a monolithic member a composite system, Cement mortar is used for these enlargements.

A later development was the use of sprayed concrete and mortar, the process referred to as shotcrete. The process was introduced in 1911 at the time when innovations in reinforced concrete technology were evolving. The widest use of section enlargement is in bridge deck rehabilitation and strengthening.

The section enlargement method is relatively easy to construct and economically effective. The disadvantages of this method are a high risk of corrosion of embedded reinforcing steel and concrete deterioration. These problems are associated with relative dimensional incompatibility between existing and new concrete. The restrained volume charges of new material are inducing tensile stresses that may lead to cracking and delaminating when the induced tensile stresses are greater than tensile strain capacity of new material.

The way to make this strengthening technique effective in the future is to use materials with higher tensile strain capacity, with low shrinkage properties.

The use of jacketing is primarily applicable to the repair of deteriorated columns, piers and piles. It is especially useful where all or a portion of the section to be repaired is under water. Jacketing consists of restoring or increasing the section of an existing member, principally a compression member, by encasement in new concrete. This method is applicable for protecting a section against further deterioration as well as strengthening. The form for the jacket should be provided with spacers to assure clearance between it and the existing concrete surface. The form may be temporary or permanent and may consist of timber, wrought iron, precast concrete or gauge metal, depending on the purpose and exposure. For marine environments or elsewhere where it is desired to protect concrete from chemical reaction with its environment or from weathering, the use of permanent timber forms is recommended; provided the appearance of the form is not objectionable and does not constitute a fire hazard. Wrought iron makes a very satisfactory, permanent form, but is expensive and so is limited to installations where the additional cost is justified by the increased life of the repair or where it is desired to protect the concrete from severe abrasion, such as due to heavy masses of moving ice. Gauge metal and other temporary forms can also be used under certain conditions.

Filling up the forms can be done by pumping the grout, by using pre packed concrete, by using a tremie, or, for subaqueous works, by dewatering the form and placing the concrete in the dry. Filling the form by pumping the grout, offers an advantage compared to others, because reliable results can be obtained with less dependence upon the skill of workmen. The use of a grout having a cement sand ratio by volume, between 1:2 and 1:3, is recommended. The richer grout is preferred for thinner sections and the leaner mixture for heavier sections. The grout should be placed as soon as possible after the rinsing operation and under an air pressure, sufficient to assure a smooth and continuous flow. The forms should be filled to overflowing, the grout allowed to settle for about 20 minutes, and the forms refilled to overflowing. The outside of the forms should be vibrated during placing of the grout. The top of the jacket should be finished with a collar of concrete in such a manner that a smooth transition between repaired and existing work will result.

In this method the entire height of the column section is increased and a cage of additional main reinforcement bars with shear stirrups is provided right from the foundation as per the requirement of additional load, etc. However, there are many instances where the column section is increased with additional reinforcement bars only on one face, and that too starting from the floor slab level of a particular floor and only up to the height of deterioration of the column. This additional section with additional reinforcement helps by reducing the floor space of the room with obstacles. This is commonly referred to as jacketing.

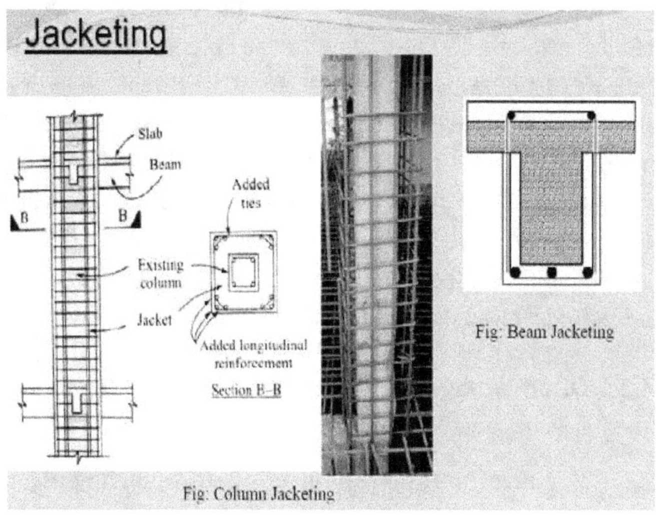

Figure 4.1 : jacketing on beam and column

4.2 Post Tensioning

External prestressing techniques have been employed with great success to correct excessive and undesirable deflections in existing structures. They have also been used to strengthen existing concrete structures to carry additional loads. Prestressing may be used on tile inside of box girders or the outside of I girders to increase the capacity of existing bridges and to provide improved resistance to fatigue and cracking.

4.3 CFRP (Carbon Fiber Reinforced Plastics) For Repair And Strengthening

CFRP has high strength, excellent strength to weight ratio, resistant to chemicals (acids and bases), good fatigue strength, and nonmagnetic, non-corrosive and nonconductive properties. As with any composite system, bond of the strengthening plates to the existing concrete is very critical. Therefore, the surface preparation of both phases of tile system, concrete and CRFP plates is very important. The plates should be ground on tile bonding side, immediately before bonding; the surface should be cleaned with acetone. After mixing, the epoxy glue component should be placed oil tile plate without delay, after assembling the plate in the designed position, a slight pressure is applied to squeeze out excessive adhesive.

4.3.1 Disadvantages Of CFRP

1. Lack of codes of practice and design standards.

2. Limitation in application to certain geometrical shapes.

3. Necessity of personnel skilled in polymers.

4. Fire protection required. Materials are being used in most cases to repair and strengthen concrete beams and columns.

   The latest studies and experience demonstrate that repairs of concrete slabs require less FRP material to achieve equivalent increase in stiffness and strength compared with reinforced concrete beams.

4.4 Materials For Repair

The basic choice of repair system is between those based on Portland cement and those on synthetic resins. In reinforced concrete, they protect the reinforcement from corrosion in different ways. Cement based materials provide an alkaline environment for the steel (pH of order of 12) and, in these conditions a passivating film forms on the surface of steel. Corrosion will occur if the alkalinity of concrete surrounding the steel is reduced by carbonation – i.e. the penetration of carbon from atmosphere or if aggressive ions such as chlorides are present. Consequently, the provision of adequate thickness of dense concrete cover is important. Resin based materials do not generally provide an alkaline environment, they normally for their protective effect on providing cover that will exclude oxygen and moisture, without which corrosion will not take place.

It is usually desirable that the mechanical properties of the repair materials should resemble as closely as possible those of the structures being repaired. This means that, as a general rule, careful consideration should be given to the use of cement based repairs. They can be made relatively inconspicuous although it is very difficult to hide them altogether without using an overall coating. They can be provided fire resistance, while resins soften at relatively low temperatures. Cement is cheaper than resin but this is seldom the deciding factor because labour usually accounts for a large proportion of repair costs. Most building operatives are more familiar with the use of cement-based material than with resin.

For some applications, however resins are more suitable. Their properties can be adjusted within fairly wide margins by suitable formulation so that they can, to some extent, be tailored to fit the job in hand. This is particularly valuable when working time is limited and rapid curing is required. Sometimes the thickness of cover to reinforcement is less than it should be and it moves the reinforcement. In this cases resin mortars may provide less permeable cover than cement mortar although the permeability can be reduced by avoided and, although polymer admixtures may make it possible to use cement mortar patches, resin mortars are often more suitable. Some compounds are not suitable for use in confined spaces and good ventilation is always desirable.

4.4.1 Polymer Modified Concrete/Cement Mortar

Polymer cements concrete, which is prepared by adding polymer or monomer to ordinary fresh cement concrete during mixing. This is based on first hand experiences of repair and restoration works of high rise building, bridges, marine installations and bomb-blast affected structures.

4.4.2 Fiber-Reinforced Plastics

These materials that are used for cracks are applied over it like a patch, using high strength epoxy adhesive increasing their service life and fortify steel or concrete structures against earthquakes or natural hazards.

4.4.3 Epoxy Resins

The epoxy resins are widely used in the repairing of cracks, patching and grouting of concrete, industrial flooring, structural adhesives, anti-corrosive linings, etc. Various types of resins, hardeners and modified epoxy systems are commonly used in structures.

4.4.4 Polymer-Based Latex

The structural integrity of chemically deteriorated reinforced concrete beams is restored by repairing one set of beams by epoxide resin latex and another by polymer-based latex system. It is interesting to observe an increase in the load-carrying capacity and rigidity of the beams after repair and rehabilitation work of the structure.

4.4.5 Fiber-Reinforced Polymer

Fiber-reinforced polymers or FRP's are robust materials that are highly resistant to corrosive action, have a high strength to weight ratio and are well suited for assembly line production into modular components that can be rapidly erected. However, FRP material costs are significantly greater than traditional concrete and steel materials. Therefore, cost savings due to either reduced weight, increased speed of construction or lower maintenance and increased life expectancy must offset this higher cost to make sensible use FRP materials. Fibre reinforced polymer (FRP) can be used for bridges to prevent corrosion. The FRP tube filled with concrete seems to be a good alternate to address this problem. The FRP tube can be engineered to provide sufficient confinement to filled concrete and to increase the capacity of the section in shear and compressive strength and also provide increased resistance to earthquake forces.

Because of the severe environment conditioning that bridge decks are subject to and the fact that they account for major percentage of bridge structures dead load, they are the most suitable bridge application for FRP materials. In addition, FRP decks can be constructed faster than conventional cast-in-place decks that take more time due to formwork construction, rebar placement and concrete curing. Other FRP material systems that utilize carbon or aramid fibers and epoxy resins offer superior structural performance characteristics.

4.4.6 Polymer-Based Materials

Polymer-based materials are being widely used in the building industry in various forms such as coatings, membranes, adhesives, sealants, etc because of their high durability.

4.4.7 High Performance Cement

High performance cement is the cement along with new complex admixture. High performance cement based mortars possess low permeability, high resistance to chemical attack, thermal resistance, and excellent freezing and thawing resistance. High performance cement is the cement along with new complex admixture. Intergrinding the complex admixture with clinker, gypsum, and selected mineral admixtures allows to rise the strength and durability of the cement. Strength increasing is the demand for engineering of cement with high volume of mineral admixtures or utilizing of industrial by-products and waste in the cement composition. This phenomenon allows to use a required amount of mineral admixtures like granulated blast furnace slag (35- 50%) in the composition to increase chemical and thermal resistance.

High performance cement based mortars possess low permeability, high resistance to chemical attack, thermal resistance, and excellent freezing and thawing resistance. Very low permeability of HP cement systems provides high resistance to chemical attack. HP cement system possesses excellent freezing and thawing resistance. The production and application of HP cement and concrete significantly increases the lifetime of structures and reduces cost of repairing and rehabilitation works due to high durability.

4.5 Selection Of Concrete Repair Materials

A variety of repair materials have been formulated to provide a wide range of properties. Since these properties will affect the performance of repair, selecting the correct material for specific application requires careful study. Concrete repair materials have been formulated to provide a wide range of properties. It is likely that more than one type of materials will satisfy the design criteria for durable repair of specific structure. While making selection of repair materials, it has to be seen that their chemical and mechanical properties are comparable to those of the substrate. Essential parameters for repair materials are low shrinkage properties, workability, low air and water permeability, durability. Classification based on application and Classification based on composition. In these cases other factors must be taken into consideration which includes:

A. Ease of application.

B. Cost.

C. Available labour skills and equipments.

D. Shelf life of the material.

Chapter

5

Structural Repair Based On Extent Of Damage

The structural repairs to be carried out in corrosion affected reinforced concrete structures to enhance its service life can be classified as follows:

1. Repairs to spallen concrete portions (steel and concrete)

    a. Cement based repairs

    b. Resin based repairs

2. 2. Large volume repair

    a. Poured concrete

    b. Preplaced concrete

3. 3. Sealing of cracks

    a. Cracks with no further movements expected
    b. Cracks with further movements expected

4. Surface coatings

5. Dry packing

5.1 Repair Of A Severely Corrosion Damaged Member, Where Cover Concrete Has Spallen And Reinforcement (Reduced In Cross-Sectional Area) Has Been Exposed

The repair process is started by cutting away all the loose and deteriorated concrete until the hard core is reached preferably behind the corroding reinforcement. All exposed reinforcements must be thoroughly cleaned. Loose rust or any contamination is removed by abrasive blast cleaning. Wire brushing by hand is not usually effective.

The portions of steel bars severely corroded require replacement. This is achieved by cutting away the corroded portions and replacing with new bars of the same type and size, either welded or tied to the existing bars. After the corrosion affected bars are replaced in position, immediately a protective primer (Zinc, neat resin or any other suitable coating) is applied. The primer chosen should be such that it should good adhesive strength and good adhesion to subsequent repair layers.   In order to build up the section, either cement based repair, or Resin based repair can be carried out.

5.2 Typical Repair Procedure For Corrosion Damaged Concrete

5.2.1 Cement Based Repairs

i. The slurry (bonding coat) is applied to all concrete surfaces to which bond is required and the patching mortar (readily available in pre-weighed packets) is applied, while the slurry is still tacky. (Care should be taken to wet the concrete surface before the application of the material but there must be no standing water left on the surfaces).

ii. After the prepared surfaces have been coated with bonding agent or a coating of neat cement slurry, the repair material consisting of 1:3(cement and sand) is applied in layers not exceeding 20mm thick. Each layer is to be key to receive the succeeding layers. The outer layers of cement should not be thicker than the inner layers. This is required, in order to prevent failure due to shrinkage stresses.

It should be ensured that the cement-based materials used in repairs do not dry out quickly. [48]

5.2.2 Resin Based Repairs

As usual, the priming coat is applied over the prepared surfaces to protect the surfaces. The interval between coats should not be too long; otherwise there will be bond failure.   Resin-based materials cure by exothermic chemical reaction immediately, when the constituents are mixed. It is essential that the materials should be well compacted to become impermeable, because they do not protect the steel by alkalinity.

5.2.3 Large Volume Repair

When a large volume of repair material is to be placed in members that have been extensively damaged, it becomes necessary to fix some kind of formwork and fill it with concrete or grout. The concrete is usually placed in conventional ways (poured concrete) or it may be formed by injecting grout into a mass of dry aggregate (under water work concrete).

5.2.4 Poured concrete

Defective concrete is first removed and loose concrete is chipped away from the face and around the reinforcement. Additional reinforcement can be provided by securely fastening it to the existing bars. It is necessary to protect the reinforcement by applying coating in the form of corrosion inhibiting paint like cement based polymer slurry or a resin based slurry. The formwork is so designed that the concrete fills it completely without leaving any air pockets. The joints in the formwork are sealed completely to avoid any leakage. Depending on the thickness to be poured, aggregate of maximum 20mm size (for thickness greater than 100mm) is adopted in the concrete mix, with suitable shrinkage compensating agent. In order to ensure good compaction of concrete, material vibration or external vibration using a mechanical hammer on the formwork can be imparted.

5.2.5 Preplaced Concrete

The technique is best suited for certain types of repair, particularly in under water work. In this method the formwork is erected in the normal way but it is first filled with clean specified (depending on thickness) coarse aggregate. Later cement grout is pumped into the forms from bottom until all the voids are filled as the air or water is vented at the top. It is essential that the formwork is watertight and is designed to withstand the full hydrostatic head of grout. This method offers quality concrete without segregation with minimum during shrinkage. This disadvantage is that the injected cement paste is prone to bleeding.

5.2.6 Sealing Of Cracks

Sealing of cracks by repair materials will be effective only when proper materials are injected. For this, the cause of crack has to be determined. If the cause of the crack is such that it is unlikely to recur, then it can be filled with a rigid material. But, if the crack is caused due to movement and that is likely to continue then any attempt to seal the crack against further movement may cause a new crack along the side of the old one.

5.2.6.1 Repair Of Cracks (Where No Further Movement Is Expected)

Such cracks can be sealed to prevent moisture penetration by simply brushing latex emulsion of low viscosity or cement paste containing fine quartz powder filler. The procedure for carrying out this type of repair is as follows:

The crack is thoroughly cleaned using compressed air. Superficial seal is applied over the crack at the surface by using a fast setting polyester resin or a thermoplastic material into which injection nipples are fixed at intervals. Injection is started at the lowest point and when resin reaches the next higher point, the injection gun is moved up to the next and the lower point is sealed. The process is continued until the whole crack gets sealed. The pressure used is carefully controlled to avoid bursting of the seal and concrete scale work.

5.2.6.2 Repair Of Cracks (Where Further Movement Is Expected)

When a crack is subjected to continuing movement, it is absolutely necessary to reduce the strain in it to reasonable amount. This can be easily done by widening the crack at the surface and sealing it with an elastic material such as polysulphide rubber or a performed neoprene strip.

Overlays may be used to seal the cracks and they are very useful and desirable where there are large numbers of cracks and treatment of each individual defect would be too expensive. Sealing of an active crack by use of an overlay requires that the overlay be extensible and not flexible alone. The occurrence of prolongation of a crack automatically means that there has been an elongation of the surface fibers of the concrete. Accordingly, an overlay which is flexible but not extensible, i.e. can be bent but cannot be stretched, will not seal a crack that is active. A two- or three- ply membrane of roofing felt laid in a mop coat of tar, with tar between the plies, the whole covered with a protective course of gravel, concrete or brick, functions very well for this purpose. The type of protective course depends on the use to which it will be subjected. Gravel is typically used for roofs, concrete or brick are used where fill is to be placed against the overlay. An asphalt block pavement also works well where the area is subjected to heavy traffic.

In addition to seal cracks, an overlay may also be used to restore a spalled or disintegrated surface. Overlays used include mortar, bituminous compounds, and epoxies. They should be bonded to the existing concrete surface.

5.2.7 Surface Coatings:

It is necessary, that after the completion of repair work, to treat both the repaired areas and the rest of the structure with some coatings, principally, to reduce the permeability of concrete, to moisture, carbon dioxide, and other aggressive agents. The coatings further can also give aesthetic look to the structure by containing the patches, discolouration and stains and match colour and textures.

Several coatings are available in the market, which can be readily used on the repaired surfaces as per the instructions of the manufacturer. Siloxene based coatings prove to be effective.

5.2.8 Dry Packing:

Dry packing or plugging is the hand placement of a low w/c ratio mortar followed by ramming or tamping of mortar into place producing an intimate contact between new and existing work. The method is applicable to cracks in a structure. Shrinkage is considerably reduced. Provides good strength and water tightness increasing the durability. Care is to be taken to use well-graded sand in the mortar mix. Drypacking is the hand placement of a very dry mortar and the subsequent tamping of the mortar into place, producing an intimate contact between the new and existing works. Because of the low water-cement ratio of the material, there is little shrinkage, and the patch remains tight. So it will be of good quality with respect to durability, strength and water tightness. Drypacking is used for filling small, relatively deep holes, such as those resulting from the removal of form ties, and narrow slots cut for repair of cracks.

5.2.9 Corrosion Protection With Hot-Dip Galvanizing

Hot-dip galvanized steel has been effectively used. The value of galvanizing stems from the relative corrosion resistance of zinc, which under most service conditions is considerably better than iron and steel. In addition to forming a physical barrier against corrosion, zinc, applied as a galvanized coating, cathodically protects exposed steel. Furthermore, galvanizing for protection of iron and steel is favored because of its low cost, the ease of application, and the extended maintenance-free service that it provides. Galvanizing primary component is zinc. The fundamental steps in the galvanizing process are.

5.2.9.1. Soil & Grease Removal -

A hot alkaline solution removes dirt, oil, grease, shop oil, and soluble markings. Pickling - Dilute solutions of either hydrochloric or sulfuric acid remove surface rust and mill scale to provide a chemically clean metallic surface.

5.2.9.2. Fluxing -

Steel is immersed in liquid flux (usually a zinc ammonium chloride solution) to remove oxides and to prevent oxidation prior to dipping into the molten zinc bath. In the dry galvanizing process, the item is separately dipped in a liquid flux bath, removed, allowed to dry, and then galvanized. In the wet galvanizing process, the flux floats atop the molten zinc and the item passes through the flux immediately prior to galvanizing[

5.2.9.3 Galvanizing –

it is immersed in a bath of molten zinc at between 815-850 F (435-455 o C). During galvanizing, the zinc metallurgical bonds to the steel, creating a series of highly abrasion-resistant zinc-iron alloy layers, commonly topped by a layer of impact-resistant pure zinc.

5.2.9.4 Finishing -

After the steel is withdrawn from the galvanizing bath, excess zinc is removed by draining, vibrating or - for small items - centrifuging. The galvanized item is then air-cooled or quenched in liquid.

Galvanizing is used throughout various markets to provide steel with unmatched protection from the ravages of corrosion. A wide range of steel products from nails to highway guardrail to the Brooklyn Bridges suspension wires to NASAs launch pad sound suppression system benefit from galvanizing superior corrosion prevention properties. Galvanizing delivers incredible value in terms of protecting our infrastructure. Less steel is consumed and fewer raw materials are needed because galvanizing makes bridges, roads, buildings, etc., last longer. Additionally, because galvanized steel requires no maintenance for decades, the rehabilitation against corrosion of steel is insufficient.

5.3 Repair Of Cracked Concrete:

Repairs to cracked concrete should not be embarked upon without full consideration of all the factors involved. All too often, specifies call for inappropriate or unnecessary work to be carried out because they have not given enough preliminary thought to the causes of cracking and the reason for repair.

The common reason for repairing cracked concrete is in order to prevent corrosion of reinforcement. Cracks may provide a path for ingress of carbon dioxide and/or water containing dissolved salts through the concrete cover, so it appears at first sight that they must form a corrosion hazard. Research has shown, however, that they must form a corrosion hazard. Research has shown, however, that this is not necessarily true. A number of codes specify maximum permissible crack widths for various conditions, but they do not agree with each other. A fundamental weakness of this approach lies in the fact that the crack width at the surface of the concrete will nearly always be greater than the width at the reinforcement, and the difference will depend largely on thickness of cover

Cracking at right angles to the reinforcing bar is often relatively unimportant. In this case the cracking will have an effect on the time that elapses before corrosion is initiated but it usually has little effect on its subsequent progress. Cracking along the length of bar is far more serious because a larger portion of the bar is exposed.

Crack injection may also be used to restore structural integrity. In such cases, the physical adhesion of the injection resin to the internal surface of the cracks has to be very good. This may require flushing of cracks with water in order to remove loosely adhering contaminants. It is also necessary for the resin to penetrate to the full depth of the cracks. It has been demonstrated that injection of suitable resin into cracked concrete can restore its physical properties.

5.3.1 Classification And Diagnosis

Cracks may be classified broadly as either 'Live', i.e. those where the width varies with time or 'Dead' cracks where no further movement is likely. They may also be subdivided into progressive cracks that are expected to become longer, and static cracks that are unlikely to do so. If repairs do not have to be carried out immediately, observation over a period of time will enable cracks to be classified and will assist diagnosis of the cause.

Dead load cracks are generally the result of an event that has passed, such as accidental overload, and they may usually be 'Locked' in such a way as to restore the structure as nearly as possible to its original un-cracked state. Cracks wider than about 1mm in horizontal surfaces can usually be sealed by filling them with cement grout. It must be remembered, however, that cracks often taper than the width at the reinforcement. Finer cracks and those in soffits or vertical surfaces may be sealed by injecting polymer. Epoxy resins are most frequently used when repair is being carried out in order to restore structural integrity, or when moisture is present. Cheaper polymers, a good example of which would be polyester resin, can often be used when the purpose of repair is to protect reinforcement from corrosion .in both cases the resin may be injected under gravity or positive pressure; better penetration can be achieved, however, by using vacuum assisted injection.

Cracks may be repaired in order to prevent leakage of fluids into or out of structures. Before this is done, the possibility of autogenously healing should be considered, especially if the fluids concerned in water. In many cases fine cracks are unsightly but they do not affect the durability or performance of the structure. When considering the appearance of cracks, the distance and circumstances of viewing during the service life of the structure should be taken into account. Many codes suggests that as a guide, a design maximum crack width of 0.3 mm may be acceptable, attempts to hide cracks by filling them nearly always fail, and the only really successful method is to apply some form of surface coating which usually has to be applied to the whole of the surface coating materials vary in their elasticity so it may be necessary to fill the cracks first, and the amount of subsequent movement that can be tolerated may be very small.

5.3.1.1 Cement Grout

Cracks wider than about 1mm in the upper surfaces of slabs etc. can often be sealed by brushing in dry cement followed, if necessary, by light spraying with water. This treatment will seal the upper part of cracks against ingress of moisture and carbon dioxide, but depth of penetration of cement will be variable. It will not the cracks completely but they will be less conspicuous than they would be if they were with a material not based on Portland cement. For cracks wider than 2mm it may be preferable to use cement and water grout but this is far more likely to leave marks on the surrounding concrete. Alternatively cracks caused out to a width of 5-10mm and pointed up with cement and sand mortar. Clearly, this will be more costly because of additional labour required.

5.3.1.2 Polymer Injection

When it is necessary to ensure, as far as possible that the sealant penetrates to the full depth of crack, injection of polymer grout under pressure is the method most commonly used. For relatively wide cracks that are unlikely to be blocked by debris, it may be enough to use a gravity head of few hundred millimeters, but in other cases hand-operated or mechanical pumps or pressure pots are used.

The general principal in sealing cracks by injection is to start at one end and work progressively along the crack. For cracks in vertical or inclined surfaces, injection should start at the lowest point and proceed upwards. A series of injection points are formed at intervals along the length of the crack and grout is injected into each point in turn until it starts to flow out of the next one .the point in use is then sealed off and injection is started at the next point, and so until the full length of the cracks has been treated. It has been argued that materials will travel into the concrete and along the joint to subsequent inlets at similar rates. It is for this reason that recommended intervals between injection points are normally equal to the depth of penetration required.

5.3.1.3 Injection Points

These can be formed in various ways; but it will be necessary to surface seal the cracks temporarily between them. Polymer-based materials are available for this purpose, with rapid curing properties and it is these, which are most often used. Sometimes holes are drilled into crack at intervals and injection nipples grouted in but, with normal drilling, there is some risk that the crack may become blocked by drilling dust. This risk can be reduced if hollow drills with an applied vacuum are available. Further to this, the method assumes the crack to run perpendicular to the surface, and this is not always the case.

The injection point might not necessarily connect with more suitable technique is to use flanged injection nipples that can be fixed temporarily to the concrete surface with an adhesive. Yet another method is to form gaps in the temporary surface seal at intervals along the crack and to use an injection nozzle that can be sealed adequately to the concrete by passing against the surface. This cannot be done if surface is too rough or if the crack is wide enough to allow resin to run out afterwards. Gaps in the temporarily seal can be formed by applying strips of adhesive tape across the crack, at intervals, applying the temporarily sealant to the full length, and then peeling off the strips of tape.

www.ingramcontent.com/pod-product-compliance
Lightning Source LLC
LaVergne TN
LVHW041334200726
843509LV00009B/714